Mitigating High-Impact Threats to Critical Infrastructure

EMERGING POLICY AND TECHNOLOGY

CONFERENCE PROCEEDINGS OF THE
INFRAGARD NATIONAL EMP SIG SESSIONS
AT THE 2013 DUPONT SUMMIT

Mitigating High-Impact Threats to Critical Infrastructure

EMERGING POLICY AND TECHNOLOGY

CONFERENCE PROCEEDINGS OF THE INFRAGARD NATIONAL EMP SIG SESSIONS AT THE 2013 DUPONT SUMMIT

Friday, December 6, 2013
Whittemore House
1526 New Hampshire Ave, NW
Washington, DC

EDITED BY CHARLES L. MANTO
AND ALEXANDRA A. KAEWERT

Videos, background and written materials are available at:
http://www.ipsonet.org/conferences/the-dupont-summit/infragard-videos-2013

Mitigating High-Impact Threats to Critical Infrastructure
Conference Proceedings
InfraGard National EMP SIG Sessions

Dupont Summit 2013
December 6, 2013

Whittemore House
1526 New Hampshire Avenue Northwest
Washington, DC 20036

Hosted and Published by the Policy Studies Organization

Editor-in-Chief, Charles L. Manto, Chairman of the National EMP SIG of InfraGard®
Editor, Alexandra A. Kaewert

September 2014

Table of Contents

Mr. Charles (Chuck) Manto, InfraGard National EMP SIG Chairman; Ms. Mary Lasky, Program Manager, Johns Hopkins University Applied Physics Laboratory; FBI Section Chief Peter Trahon, Cyber Division; Dr. Dane Egli, National Security Advisor, Johns Hopkins University Applied Physics Laboratory

Congressman Roscoe Bartlett, Ph.D., of Maryland;
Dr. Chris Beck, Vice President, Policy and Strategic Initiatives, Electric Infrastructure Security Council;
Dr. Peter Vincent Pry, Executive Director, National Task Force on Homeland Security;
Mr. Richard Waggel, FERC;
Honorable R. James Woolsey, Chairman of the Foundation for Defense of Democracies and a Venture Partner with Lux Capital, former Director of Central Intelligence

Dr. Frank Kesterman, Professor of Homeland Security and Cyber Security, University of Maryland University College;
Mr. Thomas MacLellan, Director, Homeland Security and Public Safety Division, National Governors Association;
Mr. Trent Teyema, FBI Washington Field Office, Assistant Special Agent in Charge;
Mr. William O. (Bill) Waddell, Director, Mission Command and Cyberspace Group, Center for Strategic Leadership and Development, U.S. Army War College; Mr. Robin Montana Williams, CWDP, Chief, National Cybersecurity Education and Awareness Branch, Department of Homeland Security

Mr. Scott McBride, National and Homeland Security, Critical Infrastructure Protection & Resilience, Idaho National Laboratory

Mr. William Murtagh, Program Coordinator, National Oceanic and Atmospheric Administration/National Weather Service Space Weather Prediction Center

Preface to the 2013 Dupont Summit
EMP SIG Proceedings

The conference proceedings of the InfraGard EMP SIG (Electromagnetic Pulse Special Interest Group) sessions at the 2013 Dupont Summit provide written presentations and background material for the video recordings available at http://www.ipsonet.org/conferences/the-dupont-summit.

This is the second year in which the EMP SIG has been able to publish the presentations of the prior year along with key events and related publications in the months following. The sessions covered high-impact threats to critical infrastructure with a special emphasis on geomagnetic disturbance (GMD), the topic of the sessions also provided by the EMP SIG at the Dupont Summits of 2012, 2011 and the contingency planning workshops and exercises with the National Defense University and the Maryland Emergency Management Agency in October 2011. This conference included analysis from NOAA of the July 23, 2012, super solar storm near miss and research on these impacts on power grids presented at a by-invitation-only session the day before by a number or organizations led by Idaho National Laboratory.

During the time of the 2013 conference, the Maine Public Utility Commission (PUC) collected public comments and filings on the topic as part of their requirement under state law LD131 to review threats to the electric power infrastructure on which the state of Maine depends. Many are included in the appendix for your convenience but can also be found by registering at the Maine PUC website at http://www.maine.gov/mpuc/online/index.shtml. Appendix researchers will notice that a number of presenters at the EMP SIG conference provided material to the Maine PUC public website. Reproducing these items in the appendix makes it possible for these conference proceedings to provide a small electronic library available by either going to the website listed or clicking through to the materials in the appendix and the videos of our past conference by using the Kindle edition. Hopefully, these resources will be of value to legislative researchers in states such as Virginia and Colorado who are preparing their own initiatives.

The EMP SIG wishes to thank the Policy Studies Organization (PSO) for its generous support of the conference and publication of these proceedings and to Ms. Alexandra Kaewert, who provided a significant amount of volunteer time editing these presentations. Of course, the EMP SIG, as a nationwide special interest group of InfraGard, appreciates the strong support of the national InfraGard board and staff, local chapters across the country, its members across the 50 states and three territories, and the Federal Bureau of Investigation who provides significant support to its InfraGard program.

These high-impact threats are difficult to discuss because of their grave nature and the resulting social and political concerns that they create. At the same time, the mitigating measures that many will be compelled to take could have very positive benefits for the sustainability of local communities nationwide and in any technology-enhanced society. Readers are encouraged to not only study the issue, but also join their community members to make our infrastructure as well as our relationships more sustainable as we address our mutual vulnerabilities.

As the chairman of the InfraGard EMP SIG and its conference session organizer, I welcome you to contact me for more information about InfraGard's EMP SIG and ways to participate in future activities. For information on InfraGard and how to join, see www.infragard.org.

Charles Leo Manto (cmanto@stop-EMP.com)
EMP SIG Chairman, InfraGard National
cmanto@stop-EMP.com

Introduction and Update

This introduction and update to high-impact threats to critical infrastructure provides links to updated studies and press articles that summarize much of what has been transpiring in the past year. It also serves as an introduction to a complete issue focused on EMP and extreme space weather that DomesticPreparedness.com plans to publish in November 2014. This article by Charles Manto can be found at this website address: http://www.domesticpreparedness.com/Infrastructure/Cyber_%26_IT/Solar_Storm_Near_Miss_%26_Threats_to_Lifeline_Infrastructure/http://www.domesticpreparedness.com/Infrastructure/Cyber_%26_IT/Solar_Storm_Near_Miss_%26_Threats_to_Lifeline_Infrastructure/ This article is reprinted with permission from the IMR Group, Inc., publisher of DomesticPreparedness.com, the DPJ Weekly Brief, and the DomPrep Journal. The IMR, Inc. offers no guarantees as to the accuracy of any information presented, but encourages all readers to use IMR, Inc. programs primarily as a resource to facilitate their own research.

Solar Storm Near Miss & Threats to Lifeline Infrastructure

CHARLES MANTO
Wednesday, September 10, 2014

In July, 2011, the InfraGard National Board and the Federal Bureau of Investigation approved the formation of the InfraGard National Electromagnetic Pulse Special Interest Group (EMP SIG) for the purpose of sharing information about threats that could affect critical infrastructure nationwide for more than a month and encouraging local communities to become more resilient. The threats specifically include manmade electromagnetic pulse (EMP), cyberattacks, coordinated physical attacks, pandemics, and extreme space weather. Many are not aware that the "100-year solar storm" creates ground-induced currents that travel up ground wires and can damage transformers and other large electronic systems that have long repair or replacement times.

High-impact threats are qualitatively different from many other threats for one main reason. Unlike hurricanes Katrina or Sandy, which affected regions and allowed other areas to rally to the aid of local communities, high-impact events have the capability of affecting much of the country simultaneously and limiting resources that are necessary for relief and recovery efforts. Instead of waiting days for help, affected regions could wait months for any meaningful aid. In a "just-in-time" society, the consequences are barely imaginable, but a historical background may help planners appreciate the need to minimize these effects.

Building National Awareness In October 2011, the National Defense University and the EMP SIG co-sponsored the first comprehensive nationwide contingency planning workshops and exercise on extreme space weather that could have a nationwide impact. Until that time, even the U.S. Department of Defense had not planned for a collapse of civilian infrastructure nationwide that would last more than a couple weeks (outside of nuclear or world war). In December 2011, the EMP SIG reported its findings in a seminar at the December 2011 Dupont Summit of the Policy Studies Organization in Washington, D.C. Less than eight months after the summit, on July 23, 2012, the earth experienced a near miss of a potentially cataclysmic solar storm.

Since then, on the first Friday of December each year, the EMP SIG has gathered top technical and policy experts to discuss such high-impact threats at subsequent Dupont Summit gatherings. Proceedings from the 2012 and 2013 summits are available online. By the second anniversary of the solar near miss, an array of scientific articles provoked attention in the international media.

On July 9, 2013, *Space Weather* published a study conducted by university and NASA researchers, entitled "A major solar eruptive event in July 2012: Defining extreme space weather scenarios." A NASA article published on July 23, 2014 quoted one of the *Space Weather* authors, Daniel Baker from the Laboratory for Atmospheric and Space Physics at the University of Colorado Boulder, "I have come away from our recent studies more convinced than ever that Earth and its inhabitants were incredibly fortunate that the 2012 eruption happened when it did…If the eruption had occurred only one week earlier, Earth would have been in the line of fire."

The NASA article cited the often-quoted 2008 National Academy of Sciences report on a FEMA-funded economic impact assessment, which stated that the total economic impact of such an event "could exceed $2 trillion or 20 times greater than the costs of a Hurricane Katrina. Multi-ton transformers damaged by such a storm might take years to repair." Baker further said, "In my view, the July 2012 storm was in all respects at least as strong as the 1859 Carrington event…The only difference is, it missed."

In another July 2014 research article published in *Space Weather*, entitled "Assessing the Impact of Space Weather on the Electric Power Grid Based on Insurance Claims for Industrial Electrical Equipment," the authors showed how even small space weather events have been causing damage to the electric power grids. Claim statistics from an examination of over 11,000 insurance claims from 2000 to 2010 revealed that "geomagnetic variability can cause malfunctions and failures in electrical and electronic devices that, in turn, lead to an estimated 500 claims per year within North America." If small events can have such an effect, it becomes a lot easier to imagine the impact of the storm that just missed Earth in 2012. In addition, this data suggests that, if protection were to be provided for equipment against the larger threat, then money would be saved on a day-to-day basis for even the smaller ones.

Assessments and Studies Raising Awareness Awareness of this storm peaked when *The Washington Post* editorial board made its recommendation on 9 August 2014, "The world can and should do more to prepare, adapting satellite systems, toughening electric grids and, above all, ensuring that scientists have the tools they need to anticipate space weather…For a variety of reasons—including the threat of severely inclement space weather—lawmakers must take a wider view."

Manmade EMP poses even greater problems according to studies publicly released by the congressional EMP Commission between 2004 and 2008 and highlighted in the 14 August 2014 op-ed by R.

James Woolsey and Peter Vincent Pry, both formerly with the Central Intelligence Agency. Not only is it possible for small mobile electromagnetic interference devices to be used at relatively close range against vulnerable electronic equipment and systems, but a relatively small-yield nuclear weapon could be placed on a scud missile, launched from an offshore freighter, and detonated in the upper atmosphere (80–300 miles high) to impact multiple regions or an entire continent. The electromagnetic fields emanating from EMP weapons include those that are in the billionths of seconds—much faster than lightning strikes. They travel through the air and across any kind of conductor, particularly long power or communication wires that act as giant welcoming antennae.

A September 10, 2007 economic impact assessment by the Sage Policy Group of Baltimore showed that even a regional EMP incident between Richmond, Virginia, and Baltimore, Maryland, could cause $770 billion of economic damage, even without considering loss of equipment or secondary effects such as lack of water in a large fire. The EMP Commission gave high marks for the study methodology and results, as did the economists who did the work quoted by the Academy of Sciences. In addition, the Sage report determined that protecting even 10 percent of the most critical infrastructure could alleviate up to 60 percent of the economic losses in medium-impact scenarios.

This study shows that it can be relatively inexpensive to protect critical infrastructure and that not all infrastructure may need to be protected to the same degree. However, as in the case of extreme space weather, little has been done until now to protect civilian critical infrastructure. Numerous studies have shown that U.S. lifeline infrastructures are highly interdependent and erected much like a "house of cards." Subsequent tests by Iran of freighter-launched missiles, North Korean satellite success, and turbulence in places such as the Middle East have increased concerns about the ability of nonstate actors and the likelihood of a high-altitude nuclear EMP event.

Cyberthreats—Big and Small Cyberattacks have affected everyone, even if they have merely been an inconvenience. Fortunately, insurance and other companies have shielded communities and absorbed billions of dollars in costs resulting from effective cybercrime. The largest risks to society are likely to be experienced in the arena of industrial controls, which are largely unprotected by traditional cyberprevention techniques. Numerous reports have shown that foreign cyberattackers have already breached many utilities.

What is most telling is the public release of a Federal Energy Regulatory Commission report on March 12, 2014, which some say was for official use only, showing how the successful attack of only nine electric grid facilities could result in a nationwide power outage. The report published in *The Wall Street Journal* resulted in a hastily convened U.S. Senate hearing. There was no challenge to the accuracy of the report about the grave vulnerability the country faces, but rather only a challenge because the report was "mishandled" and leaked to the public.

Although the vast majority of cyberattacks are low-impact, high-frequency events, there is a growing concern about their ability to become high-impact, low-frequency events. Like other high-impact threats, they have the ability to cause similar levels of disaster, especially when combined with other threats. However, the right type of mitigation and preparation can reduce both the impact and the temptation for adversaries to try to use them.

What remains uncertain is the willingness to engage these high-level threats. Psychological and political views complicate the discussion—a way to impose more government regulation versus a scare tactic to raise the nation's defense and homeland security budgets. In reality, there are daily cost savings, economic development, as well as environmental and security benefits when taking a reasonable systems approach to mitigate these threats. This is especially true when local communities are more sustainable and capable of creating and managing a larger percentage of their critical power and food requirements.

Sharing the Right Information with the Right People Similar to concerns that senators have raised at past cyberthreat hearings, some may think it is a challenge to begin an EMP discussion without causing panic or providing too much information to "the bad guys." One possible solution is to engage the emergency management and contingency planner communities, who are already emotionally and intellectually accustomed to dealing with disaster planning. Another is to make better use of InfraGard. So far, InfraGard is the only federally sponsored program that requires all of its individual members to sign nondisclosure agreements so they can trust each other as they hold confidential conversations and share sensitive law enforcement information. The Federal Bureau of Investigation also provides background checks so an even greater level of trust can be achieved. These trusted and informed conversations can then lead to more-effective engagement with the public—through social media outlets—similar to the EMP SIG conferences.

This year, the EMP SIG will hold its conference on Friday, December 5, 2014. On the day before, it will conduct a by-invitation-only tabletop exercise based on a high-impact incident. For additional information or to attend the conference, visit the event page. The November 2014 issue of the *DomPrep Journal* will bring together subject matter experts to take a more in-depth look at this topic to further the EMP discussion and determine what actions may be considered to better prepare for and mitigate these threats.

Charles "Chuck" Manto is CEO of Instant Access Networks LLC, a consulting and research and development firm that produces independently tested solutions for EMP-protected microgrids and equipment shelters for telecommunications networks and data centers. He received six patents in telecommunications, in computer mass storage and EMP protection and has another one pending for a smart microgrid controller. He assists other entrepreneurs and investors with their intellectual property strategies and has developed valuation methodology accepted by the U.S. Department of Defense, countries, and companies participating in industrial defense conversion. He is a senior member of the IEEE and founded and leads InfraGard National's EMP SIG. He can be reached at cmanto@stop-EMP.com.

INL Idaho National Laboratory

GEOMAGNETIC DISTURBANCE WORKSHOP

Policy Studies Organization
1527 New Hampshire Ave., NW, Washington, DC 20036
Auditorium

Thursday, December 5, 2013

12:00	Welcome	Mr. Chuck Manto InfraGard National EMP SIG Chair
	Introduction	Congressman Trent Franks 8th District of Arizona
12:30	Keynote Presentation-Space Weather	Mr. Bill Murtagh NOAA Space Weather Prediction Center
1:15	DTRA MHD-E3 Phase IVB, Measured Harmonic Response of Power Grid Transformers Subjected to Severe E3/GIC Currents	Ms. Amber Walker SARA Inc. EM & Pulse Power Applications
2:00		*Break*
2:15	Geomagnetic Disturbance Impacts	Mr. John Kappenman, Principal Consultant Storm Analysis Consultants
2:30	Transformers & GIC	Dr. Ramsis Girgis, Technical Manager ABB St. Louis
2:45	EMPrimus/ABB Solid Ground	Mr. Gale Nordling EMPrimus, President & CEO
3:15	EMP and Space Weather Impacts	Dr. George Baker GW University/DTRA Principle Consultant
3:30	Panel—Choosing the Right Standards for EMP	Panel: Kappenman, Girgis, Nordling, Baker
4:30	Next Steps Friday and Beyond	Mr. Chuck Manto and Attendees
5:00	*Adjourn*	

INFRAGARD
MEMBERS ALLIANCE

EMP SIG ((•))
ElectroMagnetic Pulse - Special Interest Group

Dupont Summit
Science, Technology, and Environmental Policy

Mitigating High-Impact Threats to Critical Infrastructure in 2014

Historic Whittemore House, Dupont Circle
1526 New Hampshire Avenue, Washington, DC
Friday, December 6, 2013

POLICY
STUDIES
ORGANIZATION
The international association for decision makers

EMP SIG ((•))
ElectroMagnetic Pulse - Special Interest Group

National InfraGard EMP Special Interest Group

8:00–8:30   **Registration**

8:30–8:40   **Introduction**
> **Mr. Charles (Chuck) Manto,** InfraGard National EMP SIG Chairman
> **Ms. Mary Lasky**, Program Manager, John Hopkins University Applied Physics Laboratory
> **FBI Deputy Assistant Director George Piro**, Weapons of Mass Destruction Directorate

8:40–9:00   **Presentation: "Strengthening Security and Resilience in the 21st Century"**
> **Dr. Dane Egli**, National Security Advisor, Johns Hopkins University Applied Physics Laboratory

9:00–9:50   **Panel: "Federal, State and Utility Plans for Grid Protection"**
> **Honorable Roscoe Bartlett,** former U.S. Congressman
> **Dr. Chris Beck**, Vice President, Policy and Strategic Initiatives, Electric Infrastructure Security Council
> **Dr. Peter Vincent Pry**, Executive Director, National Task Force on Homeland Security
> **Mr. Trent Teyema**, FBI Washington Field Office Assistant Special Agent in Charge
> **Mr. Richard Waggel**, FERC
> **Honorable R. James Woolsey**, Chairman of the Foundation for Defense of Democracies and a Venture Partner with Lux Capital, former Director of Central Intelligence

9:50–10:20   **Presentation: "EMP Mitigation Hope: Centralized vs Decentralized Systems; Public vs Private Sectors"**
> **Honorable R. James Woolsey**, former Director of Central Intelligence

10:25–10:50   **Presentation: "Updates on Space Weather Threats for Power and Communications"**
> **Mr. William Murtagh**, Program Coordinator, National Oceanic and Atmospheric Administration/National Weather Service Space Weather Prediction Center

10:55–11:20   **Presentation: "Latest Idaho National Laboratory Research Data on GMD Impacts to Power Grid Infrastructure"**
> **Mr. Scott McBride**, National and Homeland Security, Critical Infrastructure Protection & Resilience, Idaho National Lab

11:25–12:15   **Panel:** "The FBI, DHS, DoD and the National Governors Association Public/Private Plans to Mitigate Cyber Threats against CI"
> **Dr. Frank Kesterman**, Professor of Homeland Security and Cyber Security, University of Maryland University College
> **Mr. Thomas MacLellan** , Director, Homeland Security and Public Safety Division, National Governors Association
> **Mr. Trent Teyema**, FBI Washington Field Office Assistant Special Agent in Charge
> **Mr. William O. (Bill) Waddell**, Director, Mission Command and Cyberspace Group, Center for Strategic Leadership and Development, U.S. Army War College
> **Mr. Montana Williams**, CWDP, Chief, National Cybersecurity Education & Awareness Branch Dept of Homeland Security Department of Homeland Security

12:15–12:55	**Lunch Break Procedures and Working Groups Announcement** **Ms. Mary Lasky**, Program Manager, John Hopkins University Applied Physics Laboratory
12:55–1:00	**Welcome Back/Video Clip from National Geographic's "American Blackout"**
1:00–1:20	**Presentation: "Planning for High-impact Disasters in Light of** **Recent Disasters"** **Dr. Paul Stockton**, former Assistant Secretary of Defense for Homeland Defense
1:25–2:10	**Panel: Planning for EMP and High-Impact Disasters** **Dr. Richard Andres**, Chairman, Energy Security Program, NDU **BG (NYG-Ret) Kenneth Chrosniak** **Ambassador Henry F. Cooper**, Chairman, High Frontier **MG (ret'd) Robert Newman**, U.S. Army retired and former Adjutant General of Virginia **Dr. Paul Stockton**, former Assistant Secretary of Defense for Homeland Defense **CAPT James Terbush,** USN and U.S. NORAD/NORTHCOM, J9
2:15–2:55	**Presentation: "Cost-effective Electric Power Grid Mitigation in 2014"** **Dr. William Joyce**, Chairman, Advanced Fusion Systems **Mr. George Anderson,** Founder and Chairman of the Board**,** Emprimus **Mr. Gale Nordling,** President and CEO, Emprimus
3:00–3:25	**Presentation: "Cost-effective Financing Micro-grids for Critical Infrastructure "** **Mr. Jeff Weiss**, Cofounder and Managing Director, Distributed Sun **Mr. Rahul Gupta**, PWC (invited)
3:30–3:55	**Presentation: "Cost-effective EMP Protection for Communications Networks and Power** **Sources such as First Net and Developments in EMP Protection Standards"** **Dr. George Baker**, Professor Emeritus, James Madison University **Mr. David Oppenheimer**, Pathion
4:00–4:25	**Presentation: "Energy Security Impacts on the Data Center Industry"** **Mr. Thomas Popik, Chairman, Foundation for Resilient Societies** **Dr. George Baker**, Professor Emeritus, James Madison University
4:30–4:55	**Presentation: "Maine Legislation on Power Grid Protection from Manmade and Natural** **EMP"** **Maine State Representative Andrea Boland**
5:00–5:30	**Concluding Remarks and Next Steps for EMP SIG Working Group** **Mr. Chuck Manto** **Mr. Bron Cikotas**

Policy Studies Organization
Conference Room Schedule

Held simultaneously at 1527 New Hampshire Avenue

8:30–12:15   **Overflow Auditorium Viewing** (Simulcast)

12:15–12:55   **Lunch and Networking at 1526 New Hampshire Avenue**

1:00–1:30   **Overflow Auditorium Viewing** (Simulcast)

1:30–2:55   **Emerging Persistent Communications and Surveillance Innovations for Energy Infrastructure Protection & Disaster Response—Strengthening Resilience**
         Mr. Joel Coulter (moderator), VP, Business Development, KSI Video
         Mr. Jason Barton, Founder, KSI Video, Cloud enabled Virtual Remote Sensing information, sharing/collaboration
         Mr. Kevin Baugh, Knowledge Builders
         Dr. David Bither, CEO/Founder Forward Trace, Former DoD lead for the Service Oriented Horizontal Information Exchange Model (SOHIX)
         Ms. Elysa Jones, OASIS International Progress/Standardization for the Common Alert Protocol
         Dr. Mark D. Troutman, PhD, Associate Director, Center for Infrastructure Protection/Homeland Security, George Mason University

3:00   **Room Closed**

Presenter Biographies

MR. GEORGE E. ANDERSON, FOUNDER & CHAIRMAN
Emprimus, LLC

Mr. Anderson was originally employed at the Crown Iron Works Company, a family enterprise founded in 1878. There he served as Vice President for Engineering from 1975 until his departure. He also served in the capacity of Director and Executive Vice President where he specialized in new process development, design of major equipment, safety engineering, and international sales support. Mr. Anderson left Crown in 2009 and founded Emprimus LLC, a company devoted to the production of products that defeat the effects of geomagnetically induced currents and electromagnet interference caused by solar flares, RF weapons and nuclear EMP. In addition to serving as Chairman of Emprimus, Mr. Anderson has been a 32-year member of the National Fire Protection Association and is an active member of the Committee on Solvent Extraction Plants (NFPA #36), which Mr. Anderson considers to be an excellent example of cooperation and balance between equipment manufacturers, plant engineers and operators, insurers, fire marshals, and various consultants in producing standards to reduce risk in very large systems handling flammable solvents in an extraction process – informed, flexible, performance-based, near-consensus standards accepted with little resistance or friction in many nations worldwide. Mr. Anderson has an Engineering degree from Stanford University, 1969. He also served as President of the Stanford American Society of Mechanical Engineers student organization, 1968. Mr. Anderson can be reached at ganderson@crowniron.com.

DR. RICHARD ANDRES
Chairman, Energy Security Program, National Defense University

Dr. Andres is Chairman of the Energy Security Program at the National Defense University at Ft. McNair, Washington, DC. His current work focuses on energy and environmental security and particularly defense-related energy issues. Prior to joining INSS, Dr. Andres was a professor at Air University assigned to the Pentagon where he served as Special Advisor to the Secretary of the Air Force. He has also served as a consultant to the Office of the Secretary of Defense (during both the Clinton and Bush administrations), the Joint Chiefs of Staff, the Office of Force Transformation, U.S. Strategic Command, the Nuclear Posture Review, the Council on Foreign Relations and other organizations. His publications appear in such journals as *International Security*, the *Journal of Strategic Studies*, *Security Studies*, and *Joint Force Quarterly*. Dr. Andres was awarded the medal for Meritorious Civilian Service, and has received numerous academic awards and fellowships. He received his PhD from the University of California, Davis. Sample articles include:

"Volatility in the European Energy Security Framework: Addressing Ukraine-Russia Gas Pricing Disputes," *INSS Strategic Forum* (2nd Quarter 2010). "Small Nuclear Reactors for Military Installations: Capabilities, Costs, and Technological Implications," *INSS Strategic Forum* (2nd Quarter 2010).

"Energy and Environmental Insecurity," Joint Forces Quarterly, Vol 55 (4th Quarter 2009). "The Emerging Energy Security System," book chapter, in Global Strategic Assessment (National Defense University Press 2009). "The Department of Defense: New Energy Infrastructure and Fuels," book chapter, in Global Strategic Assessment (National Defense University Press 2009). Dr. Andres can be reached at rich.andres@gc.ndu.edu.

DR. GEORGE BAKER
Emeritus Professor of Applied Science, James Madison University

Dr. Baker is emeritus professor of applied science at James Madison University (JMU). In addition to teaching graduate and undergraduate S&T courses at JMU, he directed the start-up and served as Technical Director of the university's Institute for Infrastructure and Information Assurance (IIIA). Much of his career was spent at the Defense Nuclear Agency (DNA) and the Defense Threat Reduction Agency (DTRA) protecting strategic systems against electromagnetic pulse (EMP) and developing protection guidelines and standards.

He led DNA's EMP research and development program during 1987–1994 and recently served as principal staff for the Congressional EMP Commission. A primary research interest stems from his experience as Director, Springfield Research Facility—a national center for critical system vulnerability assessment. He applies lessons-learned from DoD experience to critical national infrastructure assurance and community resilience. He consults in the areas of critical infrastructure protection, EMP and geomagnetic disturbance (GMD) protection, nuclear and directed energy weapon effects, and risk assessment. He presently serves on the Board of Directors of the Foundation for Resilient Societies, the Board of Advisors for the Congressional Task Force on National and Homeland Security, the JMU Research and Public Service Advisory Board, and the National Defense Industrial Association (NDIA) Homeland Security Executive Board. Professor Baker can be reached at **bakergh@jmu.edu**.

HONORABLE DR. ROSCOE G. BARTLETT
Former Congressman for the 6th Congressional District

Elected to serve his tenth term in the United States House of Representatives, Roscoe G. Bartlett considers himself a citizen-legislator, not a politician. Prior to his election to Congress, he pursued successful careers as a professor, research scientist and inventor, small business owner, and farmer. He was first elected in 1992 to represent Maryland's Sixth District. In the 112th Congress, Bartlett serves as Chairman of the Tactical Air and Land Forces Subcommittee of the House Armed Services Committee. Owing to his 10 years of experience as a small business owner, he also serves on the Small Business Committee. One of three scientists in the Congress, Dr. Bartlett is also a senior member of the Science, Space and Technology Committee. Prior to his election to the Congress, Dr. Bartlett worked for more than 20 years as a scientist and engineer on research and development programs for the military and NASA. Nineteen of his 20 patents are held by the U.S. Government for his inventions of life support equipment used by military pilots, astronauts, search and rescue personnel, and firefighters.

In 2008, *Slate* magazine applauded him as "an advocate for reducing dependency on fossil fuels." The Association for the Study of Peak Oil (ASPO-USA) created the Roscoe G. Bartlett "Speak Truth to Power" Award in his honor in 2008. It had previously awarded him the M. King Hubbert Award in 2006 for his leadership in the Congress to promote efficiency and conservation and alternative renewable sources of domestic energy to enable the United States to overcome the challenges to national security and economic prosperity of global peak oil. Congressman Bartlett is the cofounder and cochairman of the Congressional Peak Oil Caucus. He is also the cochairman of the House Renewable Energy and Energy Efficiency Caucus and Defense Energy Security Caucus. He is also a member of the Oil and National Security Caucus. Dr. Bartlett can be reached at **roscoegbartlett@gmail.com**.

MR. KEVIN A. BAUGH
President, Knowledge Bridge International, Inc.

Kevin Baugh began his career as a Naval Special Warfare Officer, serving with distinction both as a leader and as advisor to key military and civilian decision-makers within the Department of Defense. Following his retirement from active military service, he served as the Assistant for Operations Research within the Office of the Assistant Secretary of Defense For Special Operations and Low-Intensity conflict. Since leaving government service, Kevin has provided highly specialized consulting services to the U.S. Government with a particular focus on the application of emerging technology to military problems. He currently serves as the President of Knowledge Bridge International Inc. a company focused on supporting rapidly deployable and highly scale-able communications technology solutions for industry and government.

Kevin's work has included projects assessing emerging bio-threats; developing new technologies for sensing, detection and forensics; as well as developing new approaches for sustaining human performance. Kevin is a lifetime member of the National Defense Industrial Association and the UDT/SEAL Association and the Naval Academy Alumni Association. Kevin is also serves as a volunteer with Sanctuary International, a charitable organization designed to provide support to soldiers, first responders, intelligence officers and other individuals working in high-risk positions, who are experiencing job-related post-traumatic stress. Mr. Baugh can be reached at kevin@knowledgebridgeintl.com.

DR. CHRIS BECK
Vice President, Policy & Strategic Initiatives, Electric Infrastructure Security (EIS) Council

Chris Beck is Vice President, Policy & Strategic Initiatives, of the Electric Infrastructure Security (EIS) Council. Dr. Beck is a technical and policy expert in several homeland security and national defense-related areas including critical infrastructure protection, cybersecurity, science and technology development, WMD prevention and protection, and emerging threat identification and mitigation.

Dr. Beck served as the Subcommittee Staff Director for Cybersecurity, Infrastructure Protection, and Science and Technology and was the Senior Advisor for Science and Technology for the House Committee on Homeland Security (CHS), where he worked from May 2005 to May 2011. Prior to CHS, he worked in the office of Congresswoman Loretta Sanchez for three years, beginning as a Congressional Science Fellow and then as a legislative assistant.

Before government service, Dr. Beck was a postdoctoral fellow and adjunct professor at Northeastern University. He holds a Ph.D. in physics from Tufts University (2001) and a B.S. in physics from Montana State University (1994). He served in the Marine Corps Reserve for five years (1987–1992). Dr. Beck can be reached at chris.beck@eiscouncil.org.

MR. JASON BARTON
Founder, Echostorm Inc.

Mr. Barton is a visionary and goal-oriented Senior Executive with demonstrated experience in planning, developing, and implementing cutting edge information solutions to address business opportunities. He has developed strategic plans for worldwide implementation and operation of client services, product support, quality assurance, and training. In addition, he initiated and enforced strict budget controls addressing company need and promotion of growth. Adept at crisis management, trouble-shooting, problem solving, and negotiating, Mr. Barton is also experienced in mobile application development, benchmarking, capacity planning, project and resource management.

In 2001, Mr. Barton founded Echostorm, Inc., where he served as Chief Executive Officer. Echostorm was focused on securely delivering unmanned aircraft video and associated data in near-real-time over the SIPRNET. The enterprise-class CDN was built on a Service Oriented Architecture (SOA), disseminating both live and archived video and data. Users were able to execute geospatial or keyword searches then watch video and associated data on any IP-enabled device, including iOS and Android devices. The system consisted of a patent-pending video search engine, video segmenter, CDN and specialized video encoders. Additionally, the system supported all DoD MISB approved video profiles in both Situational Awareness and Exploitation qualities. In addition, Mr. Bio served as Chief architect for the US JFCOM MAJIIC "post before process" UAV video program and co-ordinated all aspects of product launch including product development, promotions, sales, pricing and packaging. Lastly, Mr. Barton spearheaded development of a CDN and mobile app for delivery and storage of massive amounts of unmanned aircraft video feeds from the ground up, providing overall vision, business plans and project direction including creation of multiple products. Mr. Barton can be reached at jason@ksivideo.com.

MR. DAVID BITHER, MANAGING DIRECTOR
ForwardTrace, LLC

Mr. Bither has over 20 years leadership experience in business strategy and corporate and technology development within the defense, aerospace and international security industries. Currently, he is Managing Director of ForwardTrace, LLC, a company focused on strategy and portfolio development for the emerging situational intelligence and environmental sensing market. Prior to this, he served as Senior Vice President of Corporate Strategy and Development for Mav6, LLC, a systems integrator and technology development start-up specializing in design and development of analytic, situational awareness and remote sensing systems. David was responsible for strategic initiatives and corporate development for the entire Mav6 enterprise to include business capture, marketing, joint venture and M&A (mergers and acquisition) activities. Mr. Bither co-founded and chaired the Intelligence Surveillance & Reconnaissance (ISR) Technology Consortium (ISRTC), an international group of 84 industry, academic and not-for-profit organizations focused on developing the market for unmanned systems, communications and sensor payloads and data analytics to the defense, international security and commercial market sector.

Before coming to Mav6, David was a senior advisor to the Institute for Defense Analyses (IDA) and performed executive portfolio analysis with JIEDDO (Joint IED Defeat Organization). While at IDA, Dave provided threat analysis and served as AW (asymmetric warfare) ISR subject matter expert and scenario developer for DoD (Department of Defense) sponsored studies examining the department's emerging ISR architecture. At JIEDDO, he conducted project review of CIED initiative cost, schedule and performance risk and developing mitigation strategies. This included developing rapid acquisition processes for accelerating capability development and system delivery to users deployed worldwide.

Previously, Dave served as Vice President and Chief Technology Officer (CTO) with Stanley Associates, Inc. a large systems integrator headquartered in Arlington, Virginia. In this role, he managed the company's Advanced Engineering & Technology (AE&T) portfolio and oversaw the development Situational Awareness solutions for customers in the defense, homeland security and risk management sectors. Additionally, he was part of the corporate M&A team that successful acquired and integrated Morgan Research Corporation (MRC) and Techrizon, Inc. Prior to starting a business career, David was a career Army officer serving in several operational, command, and senior staff positions. These included commanding combat units during the Gulf War, Senior Software Engineer and Project Manager at PM SATCOM and as Program Manager for the DISA's Global Command and Control System (GCCS).

David obtained MS in Information Systems Technology from The George Washington University and BS degree for Regents College. He graduated from Harvard University's Executive Studies program and the Defense Acquisition University's Defense Systems Management College and is certified in both Systems Engineering and Program Management.

HONORABLE ANDREA BOLAND
State Representative, District 142, State of Maine

Rep. Andrea Boland is serving her fourth term in the House of Representatives. She serves on the Joint Standing Committee on State and Local Government and the Government Oversight Committee.

She has served on the board of the Home Health Visiting Nurses of Southern Maine, Sanford's Personnel Board, the York County Budget Committee, the Board of Directors of the Literacy Volunteers of Greater Sanford, and the Sanford Maine Stage Pinetree Players, and brought the Odyssey of the Mind program to Sanford schools. She currently serves on the Board of Governors of the National Health Federation.

Rep. Boland has been an advocate for wellness and prevention from her first year of service, arguing for implementation of nutritional health strategies into the Maine health service delivery system, including incentives and/or coverage for healthy choices and alternatives that work and providers can document. She has also introduced legislation to require doctors to give written estimates to patients for proposed procedures.

A nationally recognized leader for safe wireless and safe cell phone use and application, Rep. Boland introduced the first legislation in the world requiring health and safety warning labels on cell phones, and was instrumental in the new law directing the Maine Public Utilities Commission to examine current cyber security and privacy requirements in law relating to smart meters. This was the first legislation in Maine to address consumer concerns about smart meters and Maine has become the first state in the U.S. to allow opt-outs from wireless smart meters. Rep. Boland continues to support the effort to facilitate a good PUC response to the Maine Supreme Court's ruling that the PUC failed in its duty to examine health and safety matters before approving installation of wireless smart meters.

Rep. Boland is also considered a national leader in electric infrastructure security from geomagnetic solar storms and manmade electromagnetic pulse. She served as a U.S. delegate to the International Electric Infrastructure Security Summit in London in May of 2012, and participated in the Dupont Summit 2012 on "High Impact Threats to Critical Infrastructure" in Washington, D.C. She submitted legislation to the Leadership Council in 2012 to require known, available, low-cost protections of the grid and will be resubmitting it this session. Rep. Boland is self-employed as a real estate title examiner and as an independent distributor for a company that formulates and markets nutritional products and functional foods. Previously, she worked as a consultant to the city of Boston and successfully organized and directed an effort that returned millions of dollars to the city.

Born and raised in the Boston area, Rep. Boland graduated from Elmira College in New York with a B.A. in international studies and French. She studied for a year at the University of Paris at the Political Science Institute and earned her Master's in Business Administration from Northeastern University. In 1978 Rep. Boland moved to Maine and currently resides in Sanford. Her daughter, Michaela, lives and works in Portland and her son, Tim, lives and works in San Diego, CA. Rep. Boland can be reached at sixwings@metrocast.net.

MR. MICHAEL A. CARUSO
Director, Government & Specialty Business Development for ETS-Lindgren

He is a recognized leader in the RF Shielded Enclosure/Anechoic Chamber Industry with 30-years' experience in account management, project management, technical applications, business development, marketing and sales planning. He has participated in US and international business opportunities and projects involving, start-ups, product launches, budgeting, proposal preparation and project management His operational experience in running an EMC Laboratory adds to his depth of knowledge of real-world testing and leadership challenges.

Mr. Caruso chairs ETS-Lindgren's HEMP/EMP Product Team and has been involved in a sales, design, engineering and project management capacity for hundreds of projects involving high performance RF Shielding, both large and small over the years totaling over $75MM. Among them is the Benefield Anechoic Facility located at Edwards AFB, CA, and the very first ferrite lined 10-Meter Anechoic Chamber in North America for IBM, Austin, TX. Mr. Caruso led the EMC Power Electronics testing program for the Boeing 787-8 while at Ingenium Testing Laboratory. Mr. Caruso can be reached at Mike.Caruso@ets-lindgren.com.

BG (NYG-RET) KEN CHROSNIAK
Carlisle Fire & Rescue Company, Carlisle PA

BG (NYG-Ret) Ken Chrosniak enlisted in the Army in 1965, was later commissioned through Regular Army Officer Candidate School, and went on to complete 37 years of combined service in the Regular Army, Army Reserve, and National Guard, and was advanced to the rank of Brigadier General (New York Guard-Ret) by Governor George Pataki. He has an undergraduate degree in Secondary Education from Daemen College, and a Master's in Education from Saint Bonaventure University, and is an Army War College graduate.

Ken retired from active duty in 1998 while an instructor at the Army War College at Carlisle Barracks, where he then remained as a faculty instructor. In 2002 he was recalled to active duty service and served for two years on the Joint Chiefs of Staff, where he helped formulate the National Military Strategic Plan for the War on Terror. He then went on to serve in Iraq first at the U.S. Embassy, then later as Commander of the Abu Ghraib Forward Operating Base and detention facility. Upon release from active duty in 2005, Ken returned to Army War College instructor duties where, after one year, he was again recalled to active service for an additional two years as Chief of Staff of the Army Asymmetric Warfare Office located in the Pentagon. He presently is an Instructor in the Army War College Mission Command and Cyberspace Division.

He has served in varied command and staff assignments in the U.S. and overseas, including Vietnam, Bosnia, Kuwait, and Iraq. His most significant decorations are the Defense Superior Service Medal, the Legion of Merit, the Bronze Star Medal, the Good Conduct Medal, and the Combat Action Badge. Ken is an active Firefighter with Carlisle Fire and Rescue Company, and a member of the Carlisle Ambulance Company. However, he considers his present position within EMPact America and the EMP SIG as his most important calling in service to the Nation. Ken and his wife Gayle call Carlisle their home. His has two sons, Joshua and Christian, and a daughter Stephanie. Mr. Chrosniak can be reached at kenneth.d.chrosniak.civ@mail.mil.

DR. DANE EGLI
Senior Advisor, National Security Strategies; Johns Hopkins University, Applied Physics Laboratory

Dr. Dane Egli is a national security senior advisor at Johns Hopkins University and career Coast Guard officer who served on the White House National Security Council staff from 2004-06 as a director for counterterrorism, and as the President's advisor on hostages and global counter-narcotics. He holds master's degrees from George Washington University and National Defense University in National Security Studies, and a doctoral degree from University of Colorado in public policy. He served as the senior maritime advisor to the COCOM Commander at NORAD/USNORTHCOM from 2006-08 and speaks nationally on maritime security, national preparedness, and critical infrastructure resilience issues. Mr. Egli can be reached at dane.egli@jhuapl.edu.

MR. JOHN KAPPENMAN
Research Scientist

Mr. Kappenman has been an active researcher on geomagnetic storms, EMP, and their disruptive effects on electric power systems. He was previously employed at Minnesota Power and with Metatech Corp. He is the past chairman of the IEEE Transmission and Distribution Committee. Mr. Kappenman was a Scientific Organizer and one of the Principal Lecturers at the NATO Advanced Science Institute on Space Storms held in June 2000 and a series of other leadership roles in related activities at NOAA's Space Weather Prediction Center and NATO while also providing formal testimony for U.S. Congressional committees. He was a principal investigator on extreme space weather for the U.S. EMP Commission, FEMA under Executive Order 13407, and the 2008 National Academies of Science report on "Severe Space Weather". He has recently provided extensive information to the U.S. House Committee on Homeland Security regarding cyber, EMP, and geomagnetic storm threats to the U.S. power grid infrastructure. He is currently a member of the Joint U.S. Department of Energy/NERC Steering Committee for developing and planning a conference on High Impact Low Frequency (HILF) Threats to the U.S. Electric Power Grid, which will be held November 9–10 in Washington, DC. Mr. Kappenman can be reached at jkappenma@aol.com.

MS. MARY D. LASKY, CBCP
Program Manager, Johns Hopkins University Applied Physics Laboratory

Mary Lasky is the Program Manager for Business Continuity Planning for the Johns Hopkins University Applied Physics Laboratory (JHU/APL), and also coordinates the APL Incident Command System Team. She currently serves as Chairman of the Community Emergency Response Network (CERN) in Howard County, Maryland. She is the President of the Central Maryland Chapter of the Association of Contingency Planners (ACP). She is a member of the Nuclear/Radiation Communication Working Group.

Ms. Lasky has held a variety of supervisory positions in Information Technology and in business services. In addition, she is on the adjunct faculty of the Johns Hopkins University Whiting School of Engineering, teaching in the graduate degree program in Technical Management. She has published and presented nationally and internationally on pandemic flu, continuity of operations, and public-private partnerships. She received a MBA and a MS in Technical Management (cum laude) from Johns Hopkins University. Ms. Lasky can be reached at mary.lasky@jhuapl.edu.

MR. CHARLES MANTO
Chair, EMP-SIG

Mr. Manto is CEO of Instant Access Networks, LLC a consulting and R&D firm that produced independently tested solutions for EMP protected micro-grids. He received five patents in telecommunications and computer mass storage, has others pending (in advanced micro-grids and EMP protection), and assisted other entrepreneurs and investors with their intellectual property strategies. Developed valuation methodology accepted by the U.S. DOD, countries, and companies participating in industrial defense conversion. Facilitated due diligence of over 200 deals, managed a venture capital service, a revolving loan fund, an economic development corporation, a computer mass storage manufacturer, and broadband CLEC. Mr. Manto has also founded and leads InfraGard National's EMP SIG. He received his B.A. and M.A. from the University of IL at Urbana/Champaign and is a Senior Member of IEEE. Mr. Manto can be reached at cmanto@stop-emp.com.

MR. THOMAS MACLELLAN
Director, Homeland Security and Public Safety Division

Thomas leads the National Governors Association Center for Best Practices' work on homeland security and public safety. In this role, he is responsible for the Governors' Homeland Security Advisors Council (GHSAC), the governors' criminal justice policy advisors network, and the governors' public safety broadband advisors. Thomas has launched and directs a number of nationally significant projects across a range of policy areas including cyber security, prescription drug abuse, sentencing and corrections reform, justice information sharing, public safety broadband, as well as others. Thomas' other past projects include work on prisoner reentry, forensic DNA, juvenile justice, family violence, and school violence. Thomas holds a B.A. from the College of the Holy Cross and is a graduate of the Naval Post Graduate School Executive Leaders Program. Mr. MacLellan can be reached at tmaclellan@nga.org.

MR. SCOTT A. MCBRIDE, PE
Idaho National Laboratory

Mr. Scott A. McBride joined the Idaho National Laboratory in Idaho Falls, Idaho operated by EG&G Inc. for the U.S. Department of Energy (DOE) in 1988 with a BS degree in electrical engineering from the University of Idaho in 1987. While at the INL, Mr. McBride supported various DOE and DOD customers in numerous power systems engineering projects. In 1995 Mr. McBride was selected as the Chief Engineer for Idaho Falls Power and served in that capacity for 12 years. In 2007 Mr. McBride returned to the INL, now operated by the Battelle Energy Alliance, in the National and Homeland Security, Critical Infrastructure Protection & Resilience department. Mr. McBride is a licensed Professional Engineer in the State of Idaho. Mr. McBride can be reached at scott.mcbride2@inl.gov.

MR. DARRIN M. MYLET
Business Operations & Regulatory Affairs, Adaptrum

Mr. Mylet joined Adaptrum, a Silicon Valley-based startup exploring the applications of proprietary implementations of cognitive radio for use in wireless services and applications, in 2009. Mr. Mylet has been working globally with regulators and early adopters of white space policy and technology implementation. Mr. Mylet was with Cantor Fitzgerald in 2003-2009. Mr. Mylet worked globally with both the public and private sectors in facilitating the management and trading of (wireless) radio spectrum frequency; started the Cantor Spectrum Exchange, the first real time spectrum management and trading exchange; co-developed the business and technology platform for Cantor Gaming, the first commercial mobile wireless system at the Venetian Casino Hotel in Las Vegas in 2009; and was granted twelve patents on applying wireless to applications and business objectives.

Prior to joining Cantor-Fitzgerald, Mr. Mylet was with Radiant Networks, a U.K. based pioneer in "physical mesh" broadband wireless equipment vendor, where he was VP Sales & Marketing-Americas from 2000-2003. Prior to this position, Mr. Mylet was an executive with MFS/Worldcom/MCI from 1997 to 2000. From 1992 to 1997, Mr. Mylet was with GTE Corporation (now Verizon). Mr. Mylet served two Terms (2010/11-Obama) and (2009/10-Bush) for the Department of Commerce Spectrum Management Advisory Committee administered by NTIA and was Chairperson of the Spectrum Transparency Subcommittee. NTIA advises the White House on Spectrum and Broadband Policy.

Mr. Mylet has been serving an Advisory Role with Full Spectrum, a Menlo Park CA based company focused on Licensed Broadband Wireless for Intelligent Infrastructure using Software Define Radio since its inception in 2007. Mr. Mylet serves as an advisor to SpectrumEvolution.org. He has a B.A. in Economics Indiana University, Bloomington, IN.

MR. TOM POPIK
Chairman, Foundation for Resilient Societies

Thomas Popik is chairman of the Foundation for Resilient Societies, a nonprofit group dedicated to the protection of critical infrastructure against infrequently occurring natural and manmade disasters. He is principal author of a Petition for Rulemaking submitted to the Nuclear Regulatory Commission that would require backup power sources for spent fuel pools at nuclear power plants. Previously, as a

U.S. Air Force officer, Mr. Popik investigated unattended power systems for remote military installations. Mr. Popik graduated from MIT with a B.S. in mechanical engineering and from Harvard Business School with an M.B.A. Mr. Popik can be reached at thomasp@resilientsocieties.org.

DR. PETER VINCENT PRY
Executive Director, Task Force on National Homeland Security

Dr. Pry is the Executive Director of the Task Force on National Homeland Security. He has served: on the Commission on the Strategic Posture of the United States established by the U.S. Congress (2008–2009); as Director of the United States Nuclear Strategy Forum, an advisory body to Congress on policies to counter Weapons of Mass Destruction (2005–2009); on the Commission to Assess the Threat to the United States from Electromagnetic Pulse (EMP) Attack (also commonly known as the EMP Commission), established by the U.S. Congress (2001–2008); as Professional Staff on the House Armed Services Committee of the U.S. Congress, with portfolios in nuclear strategy, WMD, Russia, China, NATO, the Middle East, intelligence, and terrorism (1995–2001); as an Intelligence Officer with the Central Intelligence Agency responsible for analyzing Soviet and Russian nuclear strategy and operational plans (1985–1995), where he was formally recognized by the agency for his expertise, groundbreaking research, and his outstanding accomplishments during his 10 years of service; and as a Verification Analyst at the U.S. Arms Control and Disarmament Agency responsible for assessing Soviet compliance with nuclear and strategic forces arms control treaties (1984–1985).

Dr. Pry also played a key role: running hearings in Congress that warned about how terrorists and rogue states could pose an EMP threat, establishing the Congressional EMP Commission, helping the Commission develop plans to protect the United States from EMP, and working closely with senior scientists who first discovered the nuclear EMP phenomenon. Dr. Pry holds two Ph.D.s (in International Relations and U.S. History) and a certificate in nuclear weapons design from the USAF Weapons Laboratory. He has also written numerous books on national security issues. Dr. Pry can be reached at peterpry@verizon.net.

DR. PAUL N. STOCKTON
Former Assistant Secretary of Defense for Homeland Defense and Americas' Security Affairs

Paul N. Stockton was nominated by President Barack Obama to be the Assistant Secretary of Defense for Homeland Defense and Americas' Security Affairs on April 28, 2009, and was confirmed by the Senate on May 18, 2009. In this position, he is responsible for supervising the Department of Defense's homeland defense activities (including Defense Critical Infrastructure Protection and other mission assurance efforts), defense support of civil authorities, domestic crisis management, and Western Hemisphere security matters.

Assistant Secretary Stockton received a bachelor's degree from Dartmouth College Summa Cum Laude in 1976, and a doctorate in government from Harvard in 1986. From 1986 to 1989, Assistant Secretary Stockton served as legislative assistant to Senator Daniel Patrick Moynihan, advising the senator on defense, intelligence, and counter-narcotics policy, and serving as the Senator's personal representative to the Senate Foreign Relations Committee. From 1989 to 1990, Assistant Secretary Stockton was a Postdoctoral Fellow at Stanford University's Center for International Security and Cooperation.

During his graduate studies at Harvard, he served as a research associate at the International Institute for Strategic Studies in London. Dr. Stockton can be reached at pstockton@cloudpeak.sonecon.com.

CAPT JAMES W. TERBUSH, MD
USN NORAD USNORTHCOM, Science and Technology Directorate

Captain Terbush currently serves as the medical lead for Innovations and Experimentation in the Science and Technology Directorate NORAD and United States Northern Command (N-NC). N-NC Science and Technology goals are to identify and lead strategic level concepts, experiments, and technologies within N-NC, our components, DOD and other government partners, civil agencies, non-governmental organizations, academic departments and labs; and recommend innovative concepts, technologies, and relationships at senior levels to promote unity of effort in homeland defense and civil support operations to enhance N-NC capabilities, and those of our mission partners.

In 28 years of Government service, Captain Terbush has served in more than 80 countries as the Physician to forward-deployed U.S. personnel. Dr. Terbush is an honors graduate of Colorado State University and the University of Colorado, School of Health Sciences, where he earned his MD degree. He interned at Naval Hospital Camp Pendleton, CA and was trained as a Naval Flight Surgeon at the Naval Aerospace Medical Institute (NAMI) in Pensacola, FL.

During his assignment as a Flight Surgeon, he received the Navy's Flight Surgeon of the Year award for his actions with a search and rescue team in Iceland. He then completed a Family Practice residency and was Chief Resident at Naval Hospital Bremerton, WA. Following training, he was selected to be the Family Physician/Flight Surgeon at the Naval Medical Clinic London. In 1987, Captain Terbush left active Naval service to begin the private practice of medicine in rural Colorado. For the next six years, he established and managed a family practice and emergency medicine clinic. In 1991, he volunteered for and was recalled to active duty for Operations DESERT SHIELD/DESERT STORM, where he served as Casualty Receiving Officer with Fleet Hospital 6/22 in Bahrain.

In 1993, Captain Terbush returned to full-time Government service overseas. He was assigned to posts in Asia, the Middle East, the Balkans, Africa and South America as a regional physician to the U.S. Embassies in those locations. He was instrumental in bringing military operational medicine expertise to his clinical duties. He served in many austere environments including Southwest Asia. Promoted to Captain in 1993, he remained a drilling Naval Reservist during his 10-year overseas service. In 2004, Captain Terbush retired from foreign service and returned to active Naval service, where he was assigned to the Navy Medicine Office of Homeland Security. In 2005, Captain Terbush completed his Master's Degree in Public Health at the University of California, UCLA Center for Public Health and Disasters. In June 2006, he completed his Preventive Medicine and Aerospace Medicine residency at NAMI in Pensacola, FL.

From 2006 to 2009, Captain Terbush served as Command Surgeon to North American Aerospace Defense (NORAD) Command U.S. Northern Command (USNORTHCOM). In this role, he served as Principle Medical Advisor to the Commander and Staff and was responsible for the integration of DoD medical assets internally and with other agencies in support of military response to civilian disasters combating terrorism and protecting Americans. From 2009-2011, he served as the Fleet Surgeon for Commander, U.S. Naval Forces Southern Command (COMUSNAVSO) and Commander, U.S. FOURTH Fleet (C4F). With a focus on Theater Security Cooperation, COMUSNAVSO/C4F works to strengthen and build effective alliances and friendship, develop Partner Nation capabilities, and maintain U.S. operational access to defend the United States.

Captain Terbush has received numerous awards including the Defense Superior Service Medal, Meritorious Service Medal, Joint Service Commendation Medal, Navy and Marine Corps Commendation Medal, Navy and Marine Corps Achievement Medal, the Sikorsky Helicopter Rescue Award and the Mead Johnson Award for Leadership in Family Practice. He is the Immediate Past President of the American Academy of Disaster Medicine and is Board Certified in Family Practice and Disaster Medicine. Captain Terbush can be reached at jimterbush@gmail.com.

MR. RICK WAGGEL, ENGINEER
Federal Energy Regulatory Commission, Office of Electric Reliability

Rick joined the Federal Energy Regulatory Commission's - Office of Electric Reliability in 2008 where he worked until being reassigned to the Office of Energy Infrastructure Security in December of 2012. Prior to joining the Commission he spent most of his career employed with a major mid-Atlantic investor owned Utility where he gained most of his 30 plus years of experience in the electric power industry. He initially started as an engineer and gained experience with applied engineering, design engineering and standards development for the Bulk Power System. He has also been involved with system disturbance analysis, problem mitigation and transient related studies. His experience encompasses both the Transmission and Power Generation sectors with a career as a Design and Standards Engineer, Regional Engineer, Project Manager, and Project Director, where he had the responsibility of overseeing the engineering and construction of Combustion Turbine Generating Stations.

In his current capacity with the *Office of Energy Infrastructure Security* commission Rick works on physical and cyber issues for the jurisdictional energy sectors under the Commission, and is actively involved with Electromagnetic Pulse and Geomagnetic Disturbance impacts on the power grid. Rick completed his undergraduate engineering work at Grove City College and graduate work at the University of Pittsburgh.

He is also a Professional Engineer holding registry in a number of states and US Commonwealths. Mr. Waggel can be reached at Richard.waggel@ferc.gov.

MR. JEFF WEISS
Cofounder and Managing Director, Distributed Sun

Jeff Weiss is co-founder and Managing Director of Distributed Sun's Solar Energy Investment Companies (SEIC's). Mr. Weiss leads capital formation activities for SEIC's, the entities that own and operate the company's solar assets. He also plays an active management and oversight role at D-SUN. Mr. Weiss has founded, managed, and led many companies as General Manager, CFO, CMO, Board Member and venture investor. Among them are Trust Strategy Group, a $10MM strategic intelligence firm, Picture Network

International (sold to Kodak in 1997), CDx (Certificate of Deposit Exchange), and Vista Information Technologies (a $100MM network services firm). Mr. Weiss can be reached at jeff@distributedsun.com.

AMBASSADOR R. JAMES WOOLSEY
Chairman of the Foundation for Defense of Democracies; Venture Partner with Lux Capital; Former Director Central Intelligence

Mr. Woolsey previously served in the U.S. Government on five different occasions, where he held Presidential appointments in two Republican and two Democratic administrations, most recently (1993–1995) as Director of Central Intelligence. From July 2002 to March 2008 Mr. Woolsey was a Vice President and officer of Booz Allen Hamilton, and then a Venture Partner with Vantage Point Venture Partners of San Bruno, California until January 2011. He was also previously a partner at the law firm of Shea & Gardner in Washington, DC, now Goodwin Procter, where he practiced for 22 years in the fields of civil litigation, arbitration, and mediation.

During his 12 years of government service, in addition to heading the CIA and the Intelligence Community, Mr. Woolsey was: Ambassador to the Negotiation on Conventional Armed Forces in Europe (CFE), Vienna, 1989–1991; Under Secretary of the Navy, 1977–1979; and General Counsel to the U.S. Senate Committee on Armed Services, 1970–1973. He was also appointed by the President to serve on a part-time basis in Geneva, Switzerland, 1983–1986, as Delegate at Large to the U.S.–Soviet Strategic Arms Reduction Talks (START) and Nuclear and Space Arms Talks (NST). As an officer in the U.S. Army, he was an adviser on the U.S. Delegation to the Strategic Arms Limitation Talks (SALT I), Helsinki and Vienna, 1969–1970.

Mr. Woolsey serves on a range of government, corporate, and non-profit advisory boards and chairs several, including that of the Washington firm, Executive Action LLC. He serves on the National Commission on Energy Policy. He is currently Co-Chairman of the Committee on the Present Danger. He is Chairman of the Advisory Boards of the Clean Fuels Foundation and the New Uses Council, and a Trustee of the Center for Strategic & Budgetary Assessments. Previously he was Chairman of the Executive Committee of the Board of Regents of The Smithsonian Institution, and a trustee of Stanford University. He has also been a member of The National Commission on Terrorism, 1999–2000; The Commission to Assess the Ballistic Missile Threat to the U.S. (Rumsfeld Commission), 1998; The President's Commission on Federal Ethics Law Reform, 1989; The President's Blue Ribbon Commission on Defense Management (Packard Commission), 1985–1986; and The President's Commission on Strategic Forces (Scowcroft Commission), 1983.

Mr. Woolsey has served in the past as a member of boards of directors of a number of publicly and privately held companies, generally in fields related to technology and security, including Martin Marietta; British Aerospace, Inc.; Fairchild Industries; and Yurie Systems, Inc. In 2009, he was the Annenberg Distinguished Visiting Fellow at the Hoover Institution at Stanford University and in 2010–2011 he was a Senior Fellow at Yale University's Jackson Institute for Global Affairs.

Mr. Woolsey was born in Tulsa, Oklahoma, and attended Tulsa public schools, graduating from Tulsa Central High School. He received his B.A. degree from Stanford University (1963, With Great Distinction, Phi Beta Kappa), an M.A. from Oxford University (Rhodes Scholar 1963–1965), and an LL.B from Yale Law School (1968, Managing Editor of the Yale Law Journal).

Mr. Woolsey is a frequent contributor of articles to major publications, and from time to time gives public speeches and media interviews on the subjects of energy, foreign affairs, defense, and intelligence. He is married to Suzanne Haley Woolsey and they have three sons, Robert, Daniel, and Benjamin. Mr. Woolsey can be reached at jim@woolseypartners.com.

Note: Some presenters were at an alternate location during the Dupont Summit that was not part of the public sessions. Therefore, not all persons for whom biographies exist will be listed as contributors in this volume.

EMP SIG ((•))

Education. Communication. Power. Resilience.

InfraGard National EMP SIG Guidance Document

The purpose

The purpose of the InfraGard National EMP (electromagnetic pulse) SIG (special interest group) is to foster communications and coordination that will address and mitigate the threat of a simultaneous nationwide collapse of infrastructure from any hazard such as manmade or natural EMP.

Method Focusing on Local Sustainability

The National EMP SIG will mobilize subject matter experts at the national level so that local InfraGard chapters can make use of them to help local communities become more sustainable in light of these threats.

Resources

Expert National Advisory Panels

The National EMP SIG will establish panels of leading advisors. Examples may include but are not limited to:

1. a **civilian-military liaison panel** enabling local communities to become more resilient so they can better support their local military and National Guard resources;
2. a **legislative and policy liaison panel** that can facilitate discussions between interested leaders at the national and local level with those in the private sector to identify and fill policy gaps;
3. an **education panel** that can facilitate research, development, and education/training into the development of human resources needed;
4. an **investment panel** that might identify the capital support needed by local communities to enhance its sustainability;
5. a **media, communications, and outreach panel** that can facilitate communications among EMP SIG members and between other stakeholders outside the SIG;
6. an **internal coordination panel** that would coordinate activities between the EMP SIG and other SIGs and committees within InfraGard; and
7. a **communications liaison panel** that would share information about emerging communications technology and make use of it to further the activities of the EMP SIG and InfraGard.

Qualifications and Expectations of National Panel Members

Qualifications: Panel members will be chosen based on their leadership within their respective fields by virtue of knowledge, experience, roles, capabilities, or relationships.

Expectations: Panel members will agree to attend an in-person meeting once per year, several conference calls during the course of the year and occasional email or phone correspondence with EMP SIG leadership. However, given the volunteer nature of these roles, it is expected that their contributions while meaningful will be limited and offered on a best-efforts basis.

Appointment of National Advisory Panel Members: Membership to the national panels will be by appointment by the EMP SIG Chair (manager). The initial chairman of each panel, if needed, will be appointed by the EMP SIG manager and subsequently by vote of the panel members at intervals of their selection.

The InfraGard National EMP SIG leadership team includes:

1) The InfraGard National EMP SIG Chair (manager) who serves at the pleasure of the InfraGard National Members Alliance (INMA) Board of Directors.

2) The InfraGard EMP Guidance Committee composed of the INMA Chairman, the INMA President, the INMA Managing Director, and an FBI HQ Supervisory Special Agent from the National Industry Partnership Unit (NIPU) who functions as an official liaison for the FBI to InfraGard on EMP matters.

3) The InfraGard EMP SIG Senior Management Team who is appointed by the EMP SIG Chair to serve the EMP SIG membership through the following positions:

   a) The Vice-chair who is familiar with key EMP SIG activities and can take the place of the Chair during periods of the Chair's unavailability;

   b) The Administrative Officer, whose role is that of records keeper and other administrative functions of the Senior Management Team;

   c) The Strategy Officer, who will help develop and align EMP SIG policy with its activities;

   d) The Finance Officer, who will help track financial activities to support the EMP SIG and its coordination with the INMA treasurer;

   e) The Regional Outreach Facilitator, who will assist the Senior Management Team and its activities with local InfraGard chapters and Regional Reps;

   f) The Liaison Panels Facilitator, who will support the work of the Liaison Panels and their interactions with the Senior Management Team;

   g) The Subject Matter Expert Panels Facilitator, who will support the work of the SME Panels and their interaction with the Senior Management Team; and

   h) Other ad hoc committees that the Senior Management Team may deem useful from time to time.

EMP SIG Membership

Any InfraGard member in good standing may join the EMP SIG at the national level by indicating interest in participating in activities, mailings, or communications designed for membership participation. Members may either be designated as "working members" or "observers." EMP SIG members are encouraged to ask EMP SIG national leadership for help with their local EMP SIG activities.

Secure Communications between SIG Members

The InfraGard secure portal will be the primary means of communications between SIG members. EMP SIG leadership will also assist in providing other resources, including those that may be less secure, to supplement the official InfraGard communications and resources as needed on a best-efforts basis. This will include library resources and links to resources deemed to be of special value by the membership of the EMP SIG.

Other resources

The EMP SIG will be responsible for recruiting and raising resources necessary to perform its tasks subject to the normal and customary procedures and governance of the InfraGard National Members Alliance and InfraGard Members Alliances.

Governance and Activities of the EMP SIG: All activities of the EMP SIG will comply with the governance and ethics as required by the INMA Bylaws and any guidance provided by the INMA Board of Directors.

(Background: The concepts of this guidance document have been proposed by the founding SIG Chair (manager) and approved by the INMA and the FBI NIPU. See the initial authorizing letter for the EMP SIG and the initial EMP Committee/SIG proposal for additional background.)

Introductory Remarks and Opening Presentation

Introduction

Mr. Charles (Chuck) Manto, InfraGard National EMP SIG Chairman

Ms. Mary Lasky, Program Manager, Johns Hopkins University Applied Physics Laboratory

FBI Section Chief Peter Trahon, Cyber Division

Dr. Dane Egli, Johns Hopkins University Applied Physics Laboratory

"Strengthening Security and Resilience in the 21st Century"

Dr. Dane Egli, National Security Advisor, Johns Hopkins Applied Physics Laboratory

Panel: Federal, State, and Utility Plans for Grid Protection

"Federal, State, and Utility Plans for Grid Protection"

Congressman Roscoe Bartlett, Ph.D., of Maryland

Dr. Chris Beck, Vice President, Policy and Strategic Initiatives, Electric Infrastructure Security Council

Dr. Peter Vincent Pry, Executive Director, National Task Force on Homeland Security

Mr. Richard Waggel, FERC

Honorable R. James Woolsey, Chairman of the Foundation for Defense of Democracies and a Venture Partner with Lux Capital, former Director of Central Intelligence

FBI, DHS, DoD, and National Governors Association Panel

"The FBI, DHS, DoD, and the National Governors Association Public/Private Plans to Mitigate Cyber Threats against CI"

Dr. Frank Kesterman, Professor of Homeland Security and Cyber Security, University of Maryland University College

Mr. Thomas MacLellan, Director, Homeland Security and Public Safety Division, National Governors Association

Mr. Trent Teyema, FBI Washington Field Office, Assistant Special Agent in Charge

Mr. William O. (Bill) Waddell, Director, Mission Command and Cyberspace Group, Center for Strategic Leadership and Development, U.S. Army War College

Mr. Robin Montana Williams, CWDP, Chief, National Cybersecurity Education and Awareness Branch, Department of Homeland Security

Idaho National Laboratory Presentation

"Latest Idaho National Laboratory Research Data on GMD Impacts to Power Grid Infrastructure"

Mr. Scott McBride, National and Homeland Security, Critical Infrastructure Protection & Resilience, Idaho National Laboratory

Updates on Space Weather Threats for Power and Communications

Mr. William Murtagh, Program Coordinator, National Oceanic and Atmospheric Administration/National Weather Service Space Weather Prediction Center

Planning for High-Impact Disasters in Light of Recent Disasters

Major General Robert Newman, former Adjutant General of Virginia
Dr. Paul Stockton, former Assistant Secretary of Defense for Homeland Defense

Panel: Planning for EMP and High-Impact Disasters

Dr. Richard Andres, Chairman, Energy Security Program, NDU
BG (NYG-Ret) Kenneth Chrosniak
Ambassador Henry F. Cooper, Chairman, High Frontier
MG (Red) Robert Newman, U.S. Army retired and former Adjutant General of Virginia
Dr. Paul Stockton, former Assistant Secretary of Defense for Homeland Defense
CAPT James Terbush, MD, USN and U.S. NORAD/NORTHCOM, J9

Cost-effective Electric Power Grid Mitigation in 2014

Mr. George Anderson, Founder and Chairman of the Board, Emprimus
Dr. William Joyce, Chairman, President, Advanced Fusion Systems
Mr. Gale Nordling, President and CEO, Emprimus

Cost-effective Financing Micro-grids for Critical Infrastructure

Mr. Jeff Weiss, Cofounder and Managing Director, Distributed Sun

Cost-effective EMP Protection for Communications Networks and Power Sources

Dr. George Baker, Professor Emeritus, James Madison University

Mr. David Oppenheimer, Pathion

Energy Security Impacts on the Data Center Industry

Mr. Thomas Popik, Chairman, Foundation for Resilient Societies

Mr. Michael Caruso, ETS-Lindgren

Dr. George Baker, Professor Emeritus, James Madison University

Maine Legislation on Power Grid Protection from Manmade and Natural EMP

Maine State Representative Andrea Boland

Concluding Remarks and Next Steps for EMP SIG Working Group

Mr. Chuck Manto and Guests

Mr. Arnold Kishi, President of Hawaii InfraGard Members Alliance

Mr. Bron Cikotas, former executive from the Defense Nuclear Agency

Introductory Remarks and Strengthening Security and Resilience in the 21st Century

http://youtu.be/8apgVci_Fmo

Mr. Charles (Chuck) Manto, InfraGard National Electromagnetic Pulse Special Interest Group Chairman
Ms. Mary Lasky, Program Manager, Johns Hopkins University Applied Physics Laboratory
FBI Section Chief Peter Trahon, Cyber Division
Dr. Dane Egli, National Security Advisor, Johns Hopkins Applied Physics Laboratory

LAUREN SCHULER: Good morning. Thank you all for being here today. I'm thrilled to be part of this again this year. As Chuck mentioned, I'm now Supervisory Special Agent down in FBI Headquarters in the InfraGard unit, and I'm here to introduce our Section Chief in the Cyber Division, Peter Trahon, who's going to make some opening remarks from the FBI.

PETER TRAHON: Thank you Lauren, and thank you Chuck. Good morning. On behalf of the FBI, welcome to the 2013 InfraGard National EMP SIG (Electromagnetic Pulse Special Interest Group) Session. Executive Assistant Director Rick McFeeley wanted me to express his regret in not being able to be here and join you this morning. But, as an individual that helped launch the San Francisco InfraGard Chapter some 15 years ago, it's an honor and a privilege to be here in his place.

The FBI's mission is to protect and defend the United States against terrorist threats and foreign intelligence threats, as well as to uphold and enforce criminal laws. And we certainly cannot do that without our public and private partners. With over 84 member alliances and 20,000 active members, InfraGard is considered one of the most successful public/private partnerships in the federal government. InfraGard brings representatives from the private and public sectors to help protect our nation's critical infrastructures. The program is managed by the FBI Cyber Division, and each FBI field office has at least one InfraGard chapter.

For all of you that are current members, thank you for your support. For those of you that are considering being members, please join—please go to InfraGard.net. The FBI is pleased to be in partnership with the EMP Special Interest Group, which focuses on sharing information about threats, both naturally-occurring and manmade, that can have an impact on the critical infrastructures.

In reviewing today's agenda, I've noticed some very distinguished presenters. So again, welcome, and I hope you enjoy the symposium. Thank you.

CHUCK MANTO: Thank you very much. We really appreciate the support of InfraGard National's board and all the chapters around the country, and especially the FBI for their support of the program nationwide, as well. Now I would like to introduce you to Mary Lasky. She's one of our Executive

Leadership Team members, who is helping us to facilitate and support all of the working groups that we have in various subject matter, that are relevant to this. And Mary is going to have a few words so she can explain to you how you might get further involved. Mary.

MARY LASKY: Thank you, Chuck. And welcome. It's really nice to see so many of you here. We have special working groups for the EMP SIG. And the reason why we're doing this is so that we can move the agenda for protecting our country further along. We want everybody to be active and to help with this important mission. We want to make a difference, and we want you to do that too.

We have created several working groups already. And just before lunch, I will go through and talk about these in a little more detail, and introduce the people who are going to be leading that, so that you can meet them at lunchtime and talk about getting involved. But first we have the civilian military. And it's looking at such things as the National Guard, NORTHCOM (U.S. Northern Command), how do we get our citizens involved, and how do we work on that.

Then we have communication technology, and that's looking at the important kind of things: If something like this happens, how are we going to communicate? How are we going to protect our towers? How are we going to have radio communication? Cyber is another topic. Cybersecurity is really important, about how do we protect our grid from cyberattacks.

Education, on our universities and how do we get all of our people, all of the citizens of the United States to understand what is happening within EMP? And how do they need to be protected? Then we have EMP technology. And that is looking at testing, standards, protecting the grid. We also have health-related topics. And that includes not only letting the doctors know what is going on here, but looking at equipment also—medical equipment—because that is also an important topic.

Legislation and policy is another one of our things, looking at federal, state, and local issues of how do we make policy and get something really constructive done. And then space weather, which I think speaks for itself. You'll also have an opportunity for other topics that you think are really important, to specify what those are, and to let us know so that we can also have other topics, because they're so broad.

I would now like to introduce our first speaker, who is Dane Egli. Dane is a National Security Senior Advisor at the Johns Hopkins University Applied Physics Lab. He's had a very interesting career—most of it was with the Coast Guard—but he also spent 2004 to 2006 at the White House, working on counterterrorism. And then he has also spent time at NORTHCOM, and he got his PhD at the University of Colorado, but also did work at the National Defense University. It's my pleasure to introduce Dane Egli.

DANE EGLI: Thank you Mary. And good morning, everybody. Happy holidays. It's a pleasure to be with you this morning. Thank you for InfraGard, providing the leadership to allow us to convene and others as well. Today I want to talk to you about an issue that is central to this conference: How are we going to mitigate in the face of hazards that are upon us? We have a word up here, "operationalizing," and that's really the theme of what I want to cast a general context for this conference. And that is, how do we implement? How do we operationalize? How do we quantify the impact of what we're doing?

33

I'll give you a sample of who I've talked with, that will inform some of my remarks. All Hazards Consortium, working on the eastern seaboard to bring together the public and the private sector, owners and operators in particular. By the way, Google, Wal-Mart, private sector, we find that we've got to have those people in the discussion, because 60 to 85 percent of the critical infrastructure is owned and operated by the private sector.

DHS, Office of Infrastructure Protection, they are seeking ways to show impact. We have had a proliferation of policy, doctrine, guidance, and some of us have even written books in the past year. If you waterboard yourself for at least a two-year period, you might come out with something that you try and stake out a position, which that book, *Beyond the Storm*, came out about a month ago. So Office of Infrastructure Protection, they put together some of the smartest people on Planet Earth to look at these nasty issues. And then they have congressional staff and DHS leadership say, "What have you done for me? How have these millions of dollars we have invested come back in terms of operational impact—quantifiable, analytical rigor? Where is the outcome to what we've done?" We have the National Infrastructure Protection Plan (NIPP). We have the National Response Framework. We have PPD-8. We have PPD-21. We have executive orders. We have a sufficient amount of guidance, instruction, governance, on where we should go and how we should do it.

I met yesterday with Juan Hewitt, the Office of Emergency Communications, highly affected by any type of disruption that might occur in the EMP or other area. They're in the CS&C (Cybersecurity and Communications). The question that came up: How are we going to operationalize this? I was at New York University last week with Bill *Rouche?* [09:45] and about 100 people who were affected by a certain disruptive event called Sandy. They're cataloguing, inventorying, evaluating, analyzing, scrubbing, admiring, analyzing every angle of what Sandy did to the most densely populated area in America. You know what they ask? "How are we going to make a difference? How is the money that we're going to invest going to make a difference?" We invested $12 billion over 10 years before Sandy. Yet we still had water in tunnels. We had a 14-foot tidal surge. We stood up to our knees in water to get into the vehicles in the buildings that we were supposed to be using. Something has to be done differently.

I was on the phone yesterday with Austin, Texas. Texas wants to know how are we going to operationalize these great concepts. Thank you for the NIPP. Thank you for the National Response Framework. Thank you for an organization called NORAD (North American Aerospace Defense Command), and NORTHCOM. Thank you for executive orders, PPDs. Enough. What are we going to do? How are we going to show an impact? Could you show us the numbers? Oh, by the way, how are we going to optimize our resilient behavior in a resource-scarce environment? We can no longer continue to have a TSA-like response where, because people fly planes into buildings, we then form organizations where we look for knitting needles, box cutters, and fingernail clippers with an illusion of security.

I was on the phone with General *Wan?* [11:10] in Hawaii. He is the single point for Civil Defense, Homeland Security, National Guard, and he's the tag. You're it. He's it. And he's got eight islands. They have one airport with one set of airlines. They have one port with one set of cranes. And they have one oil terminal where all fuel comes in for everything they need. And then it's parsed out to the other islands. You talk about single points of failure, they get it. They understand. They have enough doctrine, policy, guidance, governance. They have had enough conferences, by the way. I literally am invited to, once a week, a conference. So there's a limit to what we can do, in terms of talking, PowerPointing, and publications.

Lastly, Israel. I spent a week in Israel recently. And they formed a new Ministry of Home Front Defense. And the question they ask in the academic IDF, as well as the new ministry, is how are we going to show an impact? Because we no longer have infinite resources. We can't just throw money at the problem, and we cannot harden our facilities enough. We cannot generate enough policy. We cannot kill enough terrorists to prevent what is inevitable.

If you're a reader, you have read 2007 Taleb, *The Black Swan: The Impact of the Highly Improbable.* No. It's not improbable, it's highly likely. What we're facing here, Steve Flynn wrote about it 10 years ago. *The Age of the Unthinkable,* [Joshua Cooper] Ramo. It's not unthinkable, it's thinkable. 2012, Zolli wrote *Resilience: Why Things Bounce Back.* No, they're not bouncing back. We have to do something different. What is that?

What is the problem? Who are the players? And how do we proceed? I'm going to go pretty quickly, because I think I have three minutes. We have 16 critical infrastructures. If an EMP occurs, or a disruption to GPS, or any one of the other ubiquitous services that we depend upon, it will affect this in a chaotic and oftentimes interdependent, cascading way. A tree can fall on a power line in Cleveland, and for the next 48 hours, because of software problems, human error, we have a cascading effect that puts the whole northeast, for 48 hours, out of power.

The takeaway for this slide, I'd like you to think about as you go into the conference, is that these are highly interdependent, yet independently governed. And the majority of these are owned and operated by the private sector. So, while we have this impulse, this fixation on generating the next document, the next policy, the next PPD, getting the words right, the fact is, we've got to get resources out where there's a quantifiable, operationalized impact that can be used by decision-makers to prioritize that which is receding, the funding that we need to secure ourselves.

So there's a dilemma here. It's a policy dilemma. It's a funding dilemma. And that is the fact that they're highly interdependent, yet managed and governed independently. Some would argue that our power grid, electricity, is about as resilient as we can afford to get it. What's that based on? Is that your instinct? Can you quantify it? Do you have a model? Do you have a framework? Okay, we're moving towards that. We've got to go there. Being from Johns Hopkins, I only get away with writing a book for one year, and putting out a report one more year. And on year three, Christine Fox, now the Deputy SecDef (secretary of defense), former CNA (Center for Naval Analyses), former CAPE (Cost Assessment and Program Evaluation), looks at us and says, "Show me the numbers. How are you quantifying this? How are you modeling this in order to make a decision?"

Okay. Why does it matter? To this audience, I don't need to dwell on this slide very long. You live this, you know this. But there are some things in there that I had to reconcile. I didn't know we had 85,000 dams. I didn't know that the majority of them are over 50 years old. And I didn't know how much these all depend in an interdependent and dependent way on all these things. If you just pick oil, and go to a single hub in Houston, go up to your tanker today at the gas station, and ask him where his truck came from, he'll say a pipeline in Virginia or Maryland or North Carolina. And, if you follow that pipeline, the majority of them, about 80 percent, go to a certain place in Texas.

What this points to is the fact that, while globalization and technology are a wonderful thing—it's given us speed, transparency, a lot of goodness—it has concurrently made us more vulnerable and more exposed to those who would do us harm. Those of us that benefit from the ubiquitous nature of this globalization's single, just-in-time delivery, global supply chains, it also makes us simultaneously more vulnerable to those elements that we're concerned with.

What's the value proposition? This is what the Israelis ask. And you'll see in *Jerusalem Post*, in the coming week, an op-ed, where we tried to wrestle this issue to the ground. It's not easy. But, if we're going to incentivize, if we're going to compel, persuade the private sector to get involved, since they own and operate 60 to 85 percent of this sector, how are we going to do it? We're going to have to show them a value proposition. We're going to have to show them a return on investment. Today, as we sit here, there's over 500 startups at the Cambridge Innovation Center between Harvard and MIT. These startups are looking for a place to invest.

Look, let's be honest. Geico no longer does mortgage insurance. Why? Because during Hurricane Andrew, they got their head handed to them. They bailed out. They do automobiles, they do motorcycles, but Geico does not do that, because there is not a profit to be made. They have to deal in terms of profits. It is not like my career, 33 years in the military, where we didn't have a profit margin. I could sometimes hide behind policy, doctrine, governance, guidance, and command and control. But, when you go to owners and operators, this is their lifeblood. If you do not have a value proposition in the resilience trade space, it is going to be a short conversation. Okay.

So here's what I would contend today. And I'm going to stop here, as a matter of fact, maybe jump to my last slide so I can get back on time. Disruptive events are inevitable. They're coming in greater numbers and greater intensity. Sandy, was it a one-off? Was Katrina an outlier? Were the Philippines an exception? Are these active shooters domestically just kind of an anomaly? No. No.

Talk to David Shedd at DIA (Defense Intelligence Agency). Talk to Michael Hayden. Talk to people who have devoted their entire life to this, other people that are here today, and they'll tell you, this is not a one-off. This is a pattern. No one government agency, no one military organization, no one private sector can deal with this. No one public/private partnership. It's going to take all hands on deck. But we've got to understand, how do we optimize in a resource-scarce environment?

The interdependencies and cascading effects can trigger disruptions anywhere in the world. A tree falling on a power line in Cleveland is one thing. But what about Fukushima? Have we distilled all the lessons learned from Fukushima, how it shut down our auto industry because of paint pigmentation? One source, just-in-time delivery, global supply chain works in our favor a lot of times. But in this case, it shut us down.

It can be something as simple as daycare centers not opening after Katrina. The shipyards in Pascagoula were ready to open, yet 50 percent of the workforce didn't show up. It wasn't about a pandemic. It wasn't about NORTHCOM putting out more doctrine. It wasn't about the White House polishing their PPD-21. It was about daycare centers not opening in Louisiana. Their workforce didn't show up because their kids were not being watched. If you follow that cascading trail, there are some very interesting implications.

We need to embed resilience as an active virtue before, during and after. A lot of people cogitate, debate. Well that's not my definition of resilience. Here is what resilience is in our research, what we've found. Anything you do, before, during or after an inevitable event, to make yourself better in the recovery, or simply to withstand what's going to happen in the future. It can be very broad and very open. That's where we have to go, in our opinion. We have to show the economic return on investment. Market competition is going to be influenced, and the areas that people are going to go to, in business and in markets and real estate, are going to be places that are not only secure, but demonstrate the attributes of resilience.

Globalization, natural disasters, demographics, terrorism, aging infrastructure. And I thought it was interesting, the Israelis say that if we shut down, physically or economically, the enemy wins, or the storm wins. So their imperative is, "Do not shut down." So, in a way, what would be an Israeli-like application? You know, we ask the question, they ask good questions. But at the end of the day, how do you operationalize that and show an impact that's quantifiable and understandable? If you can't present to a congressional committee, to a DHS leadership, to my leadership at Johns Hopkins, to General Jacoby, a quantifiable, measurable impact—not just qualitative but quantitative—we're going to be folding up our tents, I believe, and going to the next issue.

Resilience has been out there for a long time. We have to operationalize it. So, with that, I'm going to jump to this slide, which shows 11 mega communities. Over 70 percent of our GDP, over 80 percent of our population is in these 11 mega communities. How well do we understand the interdependencies and dependencies?

With that, I'm going to jump to my last slide. Oftentimes, we're given resources—say, $34 billion in UASI (Urban Areas Security Initiative) grant money—and for 10 years we decide which Humvees to buy, which proximity suits, and which cool toys. Hey, if I'm in the Port of Long Beach, I might even want underwater surveillance. I'm not sure why I need it if there is no threat assessment, maybe no analysis to support it. But it would be cool to have it. $34 billion later, we've done a lot of action.

What we would argue is that, on our way to this quantifying, operationalizing impact, how are we going to execute? We have to start to do analytics. Don't get focused on any one agent-based modeling or semantic modeling or ontology or taxonomy. We need analytics, rigorous analytics that we have, already, to go into those databases after we've risk-mapped. We have to go in and map those areas. We have 90,000 flood maps. We have soil maps for farmers to make very smart decisions on where they farm. We have canopy maps. We have navigation maps so we can keep our ships in good water. We map everything.

What we have not mapped is the terrain, peacetime, blue sky, where the dependencies reside. When we do that, we then have the ability to inform our analytics and then do action that's based on prioritization, giving decision-makers—senior decision-makers—the ability to prioritize and understand that all things are not created equal. And, if we try to protect everything, we end up protecting nothing.

Panel—Federal, State, and Utility
Plans for Grid Protection

http://youtu.be/-WBLruWHfx0

Congressman Roscoe Bartlett, Ph.D., of Maryland
Dr. Chris Beck, Vice President, Policy and Strategic Initiatives, Electric Infrastructure Security Council
Dr. Peter Vincent Pry, Executive Director, National Task Force on Homeland Security
Mr. Richard Waggel, Federal Energy Regulatory Commission
Honorable R. James Woolsey, Chairman of the Foundation for Defense of Democracies and a Venture Partner with Lux Capital, former Director of Central Intelligence

PETER VINCENT PRY: Thank you so much. Congressman Roscoe Bartlett is the reason most of us are here, actually. You know, he's the guy that held the first unclassified hearings on EMP (electromagnetic pulse). Prior to that time, it was largely classified, and he brought it out into the light of day for the public and most policymakers to know about what a dire threat this was—even though it had been known to the Department of Defense and to the intelligence community for 50 years. But he's the guy that started the congressional EMP Commission and kept it going, which came up with solutions to this, and is basically the baseline study for understanding this threat.

We owe awareness of the threat to Mr. Bartlett. And, if we ever get this country protected, and pray God we do, we'll owe that to Mr. Bartlett, too. So Mr. Bartlett, please take it away.

ROSCOE BARTLETT: Thank you. Thank you very much. It's generally agreed that there are four things that could take down the grid: a nuclear detonation above the atmosphere; a giant solar star which will occur—we may avoid the others, but we can't avoid that one—that will occur; a cyberattack; or a physical terrorist attack. And if the grid goes down, there is a general consensus that surges of electricity, as the grid goes down, will probably take out a lot of our major transformers—maybe 100, 150 or so of them.

If that happens, the grid will be down for a year or more, because we have very few spares. We don't make any of them. Those that make them have essentially no spares. So, if you need one, you order it. They will make it, and in a year, year and a half, you will have one. That's a year or more without electricity. This ends life as we know it, if you think about those 16 critical infrastructures there.

Now the innocence and ignorance on these matters in the general public is astounding. And we have a truly representative Congress. You can't imagine how difficult it was for Dr. Peter Pry and me to get the EMP Commission going. They really didn't want to do it. And we don't have time today to talk about what we had to do to get that going, but it wasn't easy in the Congress to get that thing going. By the way, we asked for the latest paper on EMP from the CIA, and they gave us Dr. Pry's paper from a decade ago. That's how little attention was being paid to this.

For 20 years I drove down the road, 50 miles down and 50 miles back every day, to Washington. And I passed I don't know how many millions of houses, some of them many, many times. And in that whole 20 years of driving back and forth, I saw one house burning—one—that was right across the road from my farm. Now, every one of those homeowners had a fire insurance policy, and yet, the probability of their home burning was very, very low. And it wouldn't end life as you know it if your home burned. It would be catastrophic for you, financially and so forth, but you'd still be here. The sun would still come up and go down, and things would be quite normal.

But, if there is an EMP attack, that won't be true, and I'm having some trouble understanding the psychology that you couldn't put your head on the pillow at night without having your home insured for fire. And yet, millions of Americans go to bed every night, and they've made essentially no preparation for an event which could end life as they know it.

What do we do? The first thing we need to do is educate: people need to understand possibility, the probability of this threat, and what the consequences would be. And then we need to have action at every conceivable level. You and your family ought to be doing something, not just some food storing. Sustainable preparation, please. You know, three months of food, that's awfully easy to store. Fifty pounds of rice at Sam's Club costs $16.44. You can store a lot of calories for very little.

But then, what do you do? What do you do if the stores don't open in three months or however long you have your food store for? You need to be sustainably prepared. You need to work with your social group, whatever it is—your club, your church—you need to work with them to be prepared. You need to work with your city or your township to be prepared. Your state needs to be prepared. And by the way, probably the most difficult level to get prepared at is the federal level.

You know, we've worked now for 20 years, Peter and I, at that level. Very, very difficult. Maybe the states can do it. Maine is already there; I understand North Carolina is going to be there shortly. But we need education. We need to be working at every one of these levels, because if this occurs—and it will occur, I shouldn't say "if," because it will occur—there will be another Carrington event, there will be another giant solar strike. If we haven't prepared, it will end life as we know it.

I can't tell you how gratifying it is to see all of you people here, because 20 years ago, there wasn't a single person in the Congress, other than Peter Pry and myself who had any interest in or concern about this subject. Thank you very much for your contribution to this. Thank you.

PETER VINCENT PRY: Well thank you, Congressman Bartlett. That's a hell of an act to follow. Thanks a lot, Chuck. Chuck has asked me to introduce myself and to give you an overview—a net assessment, as it were—of the evolving EMP threat and of the progress, or lack of progress, to protect the country...in five minutes. So I'm going to try to do that.

I'm Dr. Peter Vincent Pry—I think most of you already know me. I worked in the CIA; you've already heard from Mr. Bartlett. So enough of that. What is the bottom line, in terms of the net assessment?

Well, we have made tremendous progress, and, given how few of us, and how few resources we have to try to get this country protected, it's really amazing the progress we have made.

And yet, every year, when I deliver a net assessment, the threat is evolving faster than our efforts to buy that insurance policy that Congressman Bartlett is talking about. Start with the sun: you know, we are in the solar maximum and we could—tomorrow—get struck by a Carrington-class geomagnetic solar flare that could cause a geomagnetic superstorm.

And, in fact, there have been several such solar flares emitted by the sun during this solar maximum, including one a little over a year ago. In July 2012, there was one that narrowly crossed the path of the earth and just narrowly missed the earth. Had that hit the earth, it would have created a Carrington-class geomagnetic superstorm that would have collapsed electrical grids worldwide, and put billions of lives at risk. The congressional EMP Commission estimated that, given the nation's current state of unpreparedness, 9 of 10 Americans could die from starvation, disease and societal collapse, given a blackout that lasts a year.

What about the manmade threat? Well, in March of this year, North Korea threatened to make nuclear strikes on the United States, South Korea, and Japan. This was just four months after they demonstrated the capability to put a satellite in orbit in such a way that the height of that satellite, and the orbit that it was in, looked exactly like practice for an EMP attack. And North Korea is assessed by the EMP Commission to have a Super-EMP warhead. Just four months after they demonstrated the capability to do it, they were threatening to attack this country. It just shows how aggressive and unstable, and possibly insane, the regime's leadership is in North Korea, and how unpredictable they are.

Then, a few months after that, we intercepted that North Korean freighter trying to transit the Panama Canal, suspected of drug smuggling. We found two SA-2 nuclear-capable missiles on launchers in the hold of that freighter. Now, they didn't have warheads on them, but it was the nightmare scenario the EMP Commission predicted: that you could put a freighter off the coast of the United States, and use something like an SA-2 short-range missile to launch it, to do an EMP attack. What are they doing in the Caribbean with SA-2 missiles in the hold of a freighter? This was just in the summer of this year.

We have had incidents happen in this country—not much reported—I suspect there might have been a news blackout on it. In San Jose, California, earlier this year—I'll call them terrorists—somebody used AK-47s, did a very professional job, cut the 911 cables, and attacked the HV transformers outside of San Jose, California, and tried to destroy 15 transformers. These supported the Silicon Valley. They got away with it, because they did such a professional job of it, used AK-47s, as I said.

That was followed up by an event—this was a good news event, actually. There was an *American Blackout* National Geographic documentary. You know, I kind of put that in the positive column of things that happened. But in the negative column, something happened the very day that National Geographic ran *American Blackout*, which demonstrated very accurately what could happen in a protracted blackout by a cyberattack. The electric power industry tried to denounce the program and suggest that it was exaggerated, and yet, it was perfectly consistent with all of the major U.S. government studies that showed infrastructure collapse.

And ironically, that very morning before *American Blackout* ran, terrorists—the Knights Templar drug cartel in Mexico—deliberately blacked out a province in Mexico: put 500,000 people into blackout, so they could cut off the communications of the *federales*, go into the towns and villages, and drag out into the public square and execute about 13 of the political leaders and village leaders who were opposing the drug cartel in Mexico. That was just last month. The bad guys are already attacking electric grids and bringing them down, just across our border.

Now, what are we doing? How well are we doing? Well, the best news we've got has been mentioned already. Thank God for Andrea Boland and Sid Morris—we've got a bill passed through Maine, in three months. After trying unsuccessfully for five years to get a bill through Washington, it took Maine three months to say, "This is a serious problem. We've got to do something about it." Maine is on the way to becoming the first state in the Union to harden its electric grid against a natural and nuclear EMP, unless North Carolina beats them to the punch. Because this Tuesday, Jim Woolsey and I went down to address the North Carolina State Legislator Joint Economic Committee and Energy Committee, and they, too, want to pass an initiative.

It sort of restores my faith in democracy to see that the states get it—that the system does seem to be working at the state level. So that gives me some hope, but I don't know if it's going to be fast enough, given how fast the threats are evolving.

And in the Hill, we are still fighting at the national level. The SHIELD (Secure High-Voltage Infrastructure for Electricity from Lethal Damage) Act is not dead, but it has been stuck in the House Energy and Commerce Committee and has not been brought up for a vote. The Chairman will not let it come up for a vote. It's been stuck there for two years, even though this thing is almost identical to the GRID (Grid Reliability and Infrastructure Defense) Act, which passed unanimously when it came up for a vote in 2010, but it was stopped by one Senator.

Just before Halloween, on October 30th, Trent Franks introduced the Critical Infrastructure Protection Act—which has the support of Chairman McCall, so I expect it will pass at least the House. This Act would be a giant step forward that would introduce a 13th national planning scenario focused on EMP. All federal, state, and local emergency planning, training, and resource allocation are based on these national planning scenarios, so it's a very significant development.

Presidential Policy Directives 8 and 21 have been mentioned, and they are good policy, and are applicable to hardening the grid. But neither the President nor the Congress can require the private sector-owned utilities to harden the grid. That's why you need the SHIELD Act. This is not a sustainable or acceptable situation for our country to be in, from the point of view of this former CIA analyst and person who has always been concerned about our national security, and I hope patriot—that the survival of 9 of 10 Americans should be contingent on the proprietary interests and business interests, showing them that they can make a profit, and using moral persuasion to persuade the North American Electrical Liability Corporation to do the right thing. No, I'm a Tea Party Republican, but here is where government does need to come in and kick some butt. That's why we need the SHIELD Act. If they are not doing it—and they're not doing it—they need to be compelled to do it.

At the NCSL (National Conference of State Legislatures) yesterday, Andrea Boland and I knew the resolutions would go down to defeat, but at least we began the process of introducing resolutions to try to get the NCSL to support the SHIELD Act. It went off better than either of us thought it was going to, so there's progress there, and we will continue to work at the state level. But one of the things that's so discouraging is how the utilities don't get it. They were there in force. They were there in force in North Carolina too, agitating against this, trying to stop progress on hardening.

And here is an information sheet that was circulated, basically by the NERC (North American Electric Reliability Corporation), to educate the policymakers there on EMP, which basically says, "Don't worry about it. We're on top of it. We're going to protect the grid." And in this thing there's a statement that says, "Well, you know, we don't really have to worry about a high altitude EMP attack anyway. We don't have to worry about EMP, because if terrorists or Iran or North Korea detonated a nuclear weapon in a city, the blast and thermal effects would destroy the city and kill everybody, and the EMP is the last thing we want to know about," which shows that, just as they were oblivious, they didn't even know about that situation that had happened in Mexico when they were criticizing the National Geographic. They don't understand the fundamental fact that the kind of EMP attack we're talking about is a high altitude attack. There is no blast and thermal effects. It's only in the EMP, and ignorant armies are preventing us from getting this country protected, but they have many lawyers and deep pockets.

That concludes my remarks. Thank you.

RICHARD WAGGEL: Thank you and good morning. My name is Rick Waggel. I work for the Office of Energy Infrastructure Security for the Federal Energy Regulatory Commission (FERC), and although I'm not directly involved with reliability standards or their enforcement, I'm here today to speak on some of the efforts that have been going on at the Commission regarding geomagnetic disturbances.

And with that, I guess I have to make the disclaimer that the views I express are my own, and they don't necessarily represent those of the Commission. In May of this year, the Federal Energy Regulatory Commission issued an order requiring the development of a reliability standard to protect the bulk power supply from the impacts of geomagnetic disturbances.

Before I discuss details of the order, it's important to understand the authority that the Commission has, relative to the reliability of the bulk power system. Traditionally, the Commission's role has been in regulating the wholesale transmission rates, but in 2005, Congress enacted the Energy Policy Act, that amended the Federal Power Act to add Section 215. This section gives the Commission its authority and jurisdiction for the reliability of bulk power supply.

That section defined the bulk power system to mean the facilities, control systems, necessary to operate an interconnected electric transmission network. This also includes any generation connected to that network that's needed for the reliable operation. It doesn't include any distribution facilities.

The bulk electric system is not defined in Section 215, but it's used by the industry to define what is covered under the reliability standards, and in general, excludes distribution systems—it's the system rated 100KB and above—there are some exclusions to that.

Section 215 also authorized the commission to certify and have jurisdiction over an electric reliability organization. The purpose is to develop and enforce reliability standards to provide an adequate level of reliability for the bulk power system. In July of 2006, the Commission certified the North American Electric Reliability Corporation as the ERO (Electric Reliability Organization).

Now a reliability standard, as defined, is a requirement that provides for the reliable operation of the bulk power system. And a reliable operation, as defined in Section 215, means operating the bulk power system within limits to prevent instability, uncontrolled separation, or cascading failures as a result of a sudden disturbance or unanticipated system failure. A GMD (geomagnetic disturbance) event could be described as a sudden disturbance, and there are also indications that it does damage power system equipment.

Some things to know about the Commission's role as related to the reliability standards are that the Commission does not directly write or modify the reliability standards. That's the job of the ERO. The Commission can direct the ERO to develop a standard to meet a certain need, as was done in the case of the GMD standard. The Commission can only approve or remand a proposed standard submitted by the ERO. The Commission must approve any reliability standard before it becomes enforceable in the United States, and a standard cannot include any requirement to enlarge or construct new capacity.

The Commission's recent order is designed to protect the bulk power system from the impacts of geomagnetic disturbances, and it is meant to provide a framework that sets the high level goals of what's to be achieved by that standard. The order does not prescribe any specific methods for mitigation, and leaves the how-to up to the owners and operators. It allows the flexibility and the design on part of the owners and operators to tailor solutions to meet their specific needs.

The order does recognize the need for coordination among all entities; geomagnetic disturbances are not localized events. And finally, the goal of the GMD order is to maintain a reliable operation of the bulk power system under a severe GMD impact. The order itself was designed to allow the standards to be developed and implemented in two stages. The first stage requires the owners and operators to develop and implement operational procedures to mitigate the effects of a severe geomagnetic disturbance. These procedures are seen as an interim step until the more effective long-term solutions of stage two can be developed.

Stage one also requires that these procedures be coordinated across the region, and that the restoration plans that the utilities have account for any equipment that could be damaged or unavailable due to a geomagnetic event. The timeframe we're looking at is six months for the standard and six months, after Commission approval, for the implementation of that standard.

Stage two directs NERC to submit reliability standards to consist of two parts: an assessment phase to conduct vulnerability assessments based on the impact of a benchmark GMD event, and an implementation phase that develops and implements a mitigation plan based on the results of the vulnerability assessment of the first part.

The second stage is also to specify the severity of the GMD event and define the characteristics of the benchmark on which the systems will be assessed. The order sets some expectations, in that the standards should contain a uniform evaluation criteria, should evaluate both primary and secondary effects, and should evaluate the effects of GIC (geomagnetically induced current) on other BPS (bulk power system) equipment. So far, the focus on BPS equipment has been on transformers, but it's known to affect other equipment, such as breakers, generators, CTs (current transformers), PTs (potential transformers). It's also required to assess the impacts on a wide area regional basis, and it's to be periodically updated as changes in technology and the systems happen.

For the second stage, NERC has 18 months to draft a standard that considers a multi-phased implementation approach and prioritizes components that are vital to reliable operation of the bulk power system. With that, I'll turn it over to my partner, Dr. Chris Beck of EIS.

CHRIS BECK: Thanks Rick. My name is Chris Beck. I'm the Vice President for Policy and Strategic Commission at the Electric Infrastructure Security Council. EIS Council is a 501C nonprofit, nongovernmental organization, and our mission is to host discussions and provide education and information that's directed at electric grid resilience for electromagnetic threats that we've been discussing today: EMP, space, weather, and other threats that could damage the electric sector.

In the past, EIS's flagship activity has been hosting international summits. We've hosted four; our fifth one will be in June in London. They've bounced back and forth annually between London and Washington. They were originally attended mostly by government agencies, so there was an interest, I think, that sprung from the EMP Commission Report among legislators and government officials in both the United Kingdom and the United States. They said, this is an international problem, because all electric grids physically operate the same way. There are differences that I'll get into in a second, but the physics and the technology and the science is basically the same throughout the world.

And we'd like to have an international discussion on this. Well, if we try to go and do an official government-to-government setup of some kind of meeting like this, it'll take about 10 years. If we set up an NGO and ask them to host it, we could probably do it in 10 months. And so the EIS Council was born. We held the first summit, and we've held the three subsequent summits.

Dr. Egli this morning talked about, well, meetings are great, and summit discussions are great, and this is a great meeting as well. It's an important aspect of this issue, to get together and to share ideas and to try to broaden the number of people that know about this issue. As Peter Pry pointed out, there's still a lot of ignorance out there. The spectrum of knowledge to ignorance is complete, and the people that really understand are on the very tail of the bell curve on one end.

Through our summit series, we did notice that the coalition is broadening. Besides government, industry has become more involved. The last couple summits, we have had a very strong interest from the international insurance sector—Lloyd's of London has now commissioned two reports, and we've had a roundtable discussion as part of that, at the last summit. In fact, Lloyd's is hosting a meeting this coming January, as kind of a midterm continuation of that roundtable discussion.

And the global insurance sector is really looking at crafting products that address this risk, and how to categorize this risk, that will, we think, add a new dimension, which is an economic driver for power companies to look at as they are making decisions about whether to include GMD and EMP effects. Because to date, it just hasn't been a factor in their decision metrics. So having something, an insurance product perhaps, that says, "Well, if you look at this, and you do certain things, maybe you get a better price on your insurance. And, if you don't, maybe you don't." Now you have a new economic incentive that wasn't there before, so that's a development.

The EIS Council wants to branch out and to provide more operational type of support to the private sector and the government. And what we're doing is working on a program called the Electromagnetic Threat Protection Handbook, or E-PRO Handbook. And the idea is that it would be something where there is some real operational ability for someone to pick up this manual. It will be a living, breathing document—it's not just going to be a handbook that we send to people and they put on the shelf and forget about it—we hope.

But because this is a societal-wide issue, we can't just engage the electric power sector. If you look at the whole spectrum of before an event to an event happening to response and recovery, all the number of people that would be involved in such a widespread power outage is pretty much everybody. So we're not just talking to the electric sector, and we're not just talking to the federal government. We want to talk to state and local governments, local first responders, the NGO community, which will also be involved.

And, in a wide-scale power outage, all of those people will have actions that will build off of their traditional roles of, say, recovery from a hurricane. But it will be a different environment, the biggest challenge being that most natural disasters are somewhat localized; there is an inside of the event and an outside, and command and control is usually set up outside the event. It can evacuate people to an unaffected area. Well, in a nationwide blackout, everybody is on the inside—there is no unaffected area. And so, we just have to think about how modest changes to and supplementation of traditional activities could be effective in helping to prevent an event and better respond to it—the overall idea being increased resilience.

The first step that we took in this E-PRO package was a report that we did, called the International E-PRO Report. We worked with the Department of Energy on it. What it is is an 11-nation survey of experience with and actions taken for geomagnetic disturbances and EMP. It's not a highly technical manual. We've submitted it for the conference proceedings, and you can also find it on our website, EIScouncil.org, under Resources, so I hope you can read it. And we found some interesting things, talking about operational level stuff, that there are, in the countries we surveyed, different activities going on, some that were quite interesting and unexpected.

For example, in the hardware realm, New Zealand has a high voltage DC line that connects the North and South Islands. And they were experiencing stray DC currents that were going into the transformers on the South Island. And their engineering solution was to put resistors in the grounding path; they call them earthing resistors in New Zealand. But it diminishes—it doesn't completely block, but it diminishes—those currents going into the transformer.

Well, it turns out that this has an ancillary benefit for geomagnetic disturbances, which are also quasi-DC currents. And so, as we're looking at possible, say, technical solutions and, would grounding resistors work, well Transpower in New Zealand has 20 years of experience of using grounding resistors. So let's ask them what their experience is: How did it affect their system performance, and so on? There is some experience out there with these things.

Finland has a very tough system: they use solely a specific kind of transformer, called a three-limb core form transformer, which is noted by transformer experts as being the most resilient to quasi-DC currents, just because of its configuration. They have what are called series capacitors on all of their 400 kV transmission lines, which is their extra high voltage line. Again, the series capacitor was put in place to help fine-tune the efficiency of the transmission line, so that's the main reason that power grid operators use serious capacitors. But, because it's a capacitor, it blocks DC current, so it has an ancillary benefit.

We find examples like that. The Israelis have done a full-scale model of their power system to analyze their vulnerabilities, and they'll be looking at ways to protect their system, gathering information from these other countries. So there's some information there, and we hope that, for example, sharing information like this will help governments and power companies learn that there are solutions out there. There's no one on the planet who has it completely nailed.

Norway has a regulation in place very comparable to the one under FERC Order 779. It is completed, so they have a full-on regulation that says Statnett, which is the national transmission operator in Norway, shall protect the system from GMD and EMP. It's not very prescriptive, but it does have a requirement for them to do so. The U.S. grid is 10 times bigger and 1,000 times more complicated than any of these other grids, basically because most of the countries that we looked at have one government agency, one transmission company, one set of regulations, while the United States has thousands of companies, 50 state public utility commissions, and federal authority. And our federal authority is unique with FERC-NERC partnership that Rick just described, that you won't find anywhere else in the world, either.

So the challenge in the United States is technically and governmentally more difficult. But we believe that there are some lessons learned here, that these are good laboratories to look at—others that have had experience and are having some successes. As I said, no one is bulletproof, but in learning from each other and sharing information, we hope that all the electric power grids of the United States and our allies will be better protected. Thank you.

JIM WOOLSEY: Hi. My name is Jim Woolsey. I was Director of Central Intelligence in the Clinton administration for two years, back in the mid-'90s. I think it's important to try to get inside the heads of enemies and potential enemies of the United States in order to assess whether there really is the possibility of an EMP attack. Because one tends to, in the halls of academe and among journalists and everywhere else, say, "Oh, this is pretty theoretical. And anyway, nobody is really crazy enough to attack like that. They'd get wiped out. So why are we bothering with this?"

And the answer is, I think, twofold. First of all, the practicality of an attack: an EMP attack does not require—indeed, it is not even useful to have—a large blast, or to have accuracy on a missile, or to have high yield. None of those is necessary or even useful. An EMP attack could well take place simply by putting a small, simple nuclear weapon on a satellite, launching the satellite—if one wants to be particularly cautious—around the South Pole, a fractional orbital bombardment system, rather than over the North Pole, where all our defenses are. And then, as that weapon comes around on a satellite, say 200 miles altitude, simply to detonate it over the center of the United States. No accuracy, no requirements for particular types of reentry vehicles, shielding, none of that.

So we are talking, in the first instance, about a relatively simple process. We know it can be done, both because coronal events have triggered EMP activities, which are similar to what would occur in a nuclear blast. And also, we had atmospheric nuclear blasts at approximately the right altitude to test this concept back in 1962, right at the end of the period when we had atmospheric tests. The Russians paid more attention, I think, than we did. But there is data that suggests what would happen with, let's say, a 200 mile-high detonation.

No reasonable and persuadable (with evidence) scientists or military strategists that I know of deny this. One gets a lot of arm-waving, but the people who really know, including the Commission, do not gainsay that problem. What you tend to get is the idea that, "well, you know, the North Koreans and whoever else, Iranians, once they get a nuclear weapon, they're not crazy. They don't want to all die. And they would suffer retaliation. So let's forget about this. This is just a subcategory of nuclear deterrence."

I don't think so. In the first instance, one can conduct an EMP attack—again, around the South Pole—without giving away who one is, or where one is coming from. There is no particular reason for the North Koreans or the Iranians, once they get a nuclear weapon, to do anything to indicate who is detonating this weapon above the continental United States.

Second, people tend to say, "Well, you know, the North Koreans aren't nuts." I beg to differ, depending on your definition of nuts. In the Cuban missile crisis, we know from the Soviet database, that Castro worked very hard to try to get the Soviets to use a nuclear weapon in the Cuban missile crisis. He knew it would result in the destruction of Cuba, and he didn't care. And he was not even a religious fanatic, just a fanatic.

So if one looks at Iran today, one could have an ideology that is stimulated by some religious views, particularly extreme Shia, to the tune that it would be a good idea to have worldwide chaos and lots and lots of nuclear weapons going off, because, from their point of view, it would mean they go to heaven and we go to hell. And they each get 72 virgins, and we die.

From their point of view, both on the Iranian side and on, I think, the North Korean side, one cannot assume that, just because someone is not a raving lunatic, just because they're not ranting in the streets, but rather they think, like Castro did in '62, or they think like, perhaps, Khamenei does today. If they think that way, they could well be very interested in an EMP attack, particularly if they think they can hide where it's coming from. And they write about it all the time. This is not something that has been dreamed up. It has been subjected to very careful analysis. Peter has been involved in much of it at CIA and elsewhere.

And those people who say, "Nobody could be crazy enough to try something like this," are not students of history, and they are not people who have made a careful examination of the problem. They are people who are trying to wave their arms and wave the problem away.

I wanted to touch on this issue of possibility, of decision-making in a rogue state leading toward an actual launch of an EMP attack. One cannot, with any reasonable attention to history, or the nature such as during the Cuban missile crisis, or the nature of the weapon, come to the conclusion that this is just some fanciful subcategory of nuclear deterrence. An attack of this sort is much easier than most nuclear attacks that people talk about, which require substantial accuracy and a number of other characteristics that are not required if one is going to fire as an Iranian leader or a North Korean leader or whomever, a nuclear weapon at the United States in an EMP mode. Thank you.

CHUCK MANTO: Now we have a few minutes for questions. But what I'd like to do is have the first questions be from the panel amongst each other. You might have an interest in clarifying or asking a question of someone else in the panel first. So I just want to make certain you guys get the first shot at each other.

ROSCOE BARTLETT: I have, perhaps naively, assumed that whereas these regimes are evil, they were not suicidal. Mr. Woolsey may be correct. They may, in fact, be suicidal. But even if they are just evil and not suicidal, we still are at huge risk, because there are ways of producing an EMP where we would not know who the perpetrator was. A tramp steamer, a Scud launcher, which you can buy for about $100,000 in the open market—any crude nuclear weapon—it wouldn't take out the whole country, but it would certainly take out from Boston to Norfolk and west to Pittsburgh. That could be a fatal blow to our country.

If they missed their target by 100 miles, as Mr. Woolsey was pointing out, it wouldn't make any difference. It would still be just as effective. And then the ship is sunk; there are no fingerprints. As a matter of fact, it doesn't have to even be a country that does it. It could be a nonstate actor that does it. Since all you need is a tramp steamer, a Scud launcher, that will go 180 miles, that won't cover the whole United States, but it'll cover enough to be really, really hurtful to us.

Is this not realistic?

JIM WOOLSEY: I think it's quite a realistic threat. And the anonymity that you mentioned, and I had as well, is an important part of it. It's important to realize that during the Cuban missile crisis—and people have now interviewed the sort of squadron commander of the nuclear submarines off Cuba, the Russian nuclear submarines—we know for a fact that the Cubans and the Russians were requiring, in order to launch one of their new nuclear torpedoes, which were on a couple of submarines, in order to launch one of those, they needed three votes. One vote was from the party chairman for the submarine squadron. One vote was from the squadron commander—all three of these people were Navy captains from the Russian Navy—and the third was the ship commander of the ship that had the nuclear torpedo on it.

We know, for a fact, the commander of the submarine that had the nuclear weapon on it pushed his button and voted for that nuclear torpedo to be launched at American ships. We know that the party leader for the squadron, the Communist party leader, pushed his button and voted for the nuclear torpedo to be used. One guy stood alone—he was a Navy captain, and he was the commander of the Russian submarine—I think he was the squadron commander; the one who voted in favor of it was the submarine commander. But he was the one of the three. And he voted not to launch the nuclear torpedo, over the objections of the other two Soviet naval captains.

You cannot predict who is going to be suicidal, who is going to be merely evil, who is going to be some combination of the two, who is going to respond to culture and tradition, and who is going to respond to direct orders in a circumstance like that. But people, including one at the Kennedy School and others, who wander about conferences and so forth, and say, "Well of course, this is crazy. And nobody is going to be crazy." Like I say, they need to examine their definition of crazy, because, as Hamlet put it, it is entirely plausible to be mad north by northwest.

ROSCOE BARTLETT: I think Mr. Woolsey is correct that the thinking of many of our leaders is that the consequences of launching an attack like that would be so devastating to the person that launches it, nobody will do it. And what Mr. Woolsey is pointing out is that that may not be true. But, you know, they don't have to put themselves at risk to do a nuclear EMP, because there are many ways that they can do it and leave essentially no fingerprints on it.

JIM WOOLSEY: Weather balloon works fine.

CHUCK MANTO: I have a question, and we may have time for a couple of questions from the audience in a moment. I know that there was a lot of discussion about the importance, not only of getting federal-level development protection, regulations, legislation passed. We also talked about Maine and what can be done at the state level. And I also wanted to ask about the technology end of this. We have large regional electric grids that need EMP protection, but it seems like one of the other things we might consider is the development of local power generation and storage that could be developed all across the country, even by many of the communities and individuals represented in this room. And there may be some practical things that local communities can do long before Washington and the large utilities can do what they need to do.

So the question is have for the panel is, what are your opinions about what might be done with local power generation and storage, and the EMP protection of that?

ROSCOE BARTLETT: I was born in 1926, and I lived through the Cold War. And those of you who did that can remember all of the fallout shelters. Now bombing us was not going to end life as we know it. But we had drills—I worked at the National Institutes of Health—we had drills there until the hydrogen bomb was developed. And then we might have been in the fireballs, so we stopped having drills at the National Institutes of Health. Building 10 was going to become a large casualty hospital if that happened.

And Chuck, I've been very interested in the psychology: everybody was involved in that. We had drills at school. You had drills, you knew what you were supposed to do. That didn't even come close to the life-changing effect that an EMP attack would. Why are we, today, doing essentially nothing, compared to what we did then? There were fallout shelters with food stored. You couldn't go into a public

49

building without getting a brochure telling you what you ought to be doing, and how to do it. Why is it so darned different today?

JIM WOOLSEY: I think it's probably because of the massive cost of protecting against an EMP shot. That massive cost was laid out clearly—and I've got my tongue in my cheek, by the commission that Roscoe and Peter had so much to do with. The cost would be, to protect and put the resilience into the system that you'd need, would be approximately 20 cents for each electricity consumer in the United States.

If you go to a Starbucks and look at the coffee cup prices, a macchiato is the most expensive small coffee. And it's $3.95, whereas, for $1.95 you can get a small plain espresso without the macchiato flavoring. So that's a two-dollar difference in price. Well, if you will just, next time you're at the coffee house, if you will just sip an espresso rather than a macchiato, you can save $2. And that would be the contribution that would be required by 10 people in order to make a contribution toward protecting—a substantial contribution for those 10 people—of protecting the grid against EMP.

So, whenever somebody from the electric power industry or otherwise tells you what a big deal this is and how expensive it is, go to the report, find out where it says 20 cents per consumer, and hand them a copy.

ROSCOE BARTLETT: I think we need to put this cost in perspective. I seem to remember that we had a stimulus package that cost, what, one trillion dollars. The numbers I get is that this would cost maybe two billion dollars. I think there are a thousand billion in a trillion. So the cost of doing this is absolutely inconsequential compared to the money that we spend on other things. It has to be more than just costs that are deterring us from doing this. Because if you put this in the context of cost to prepare, it's really pretty darned small compared to the money we're spending on other things, is it not?

JIM WOOLSEY: It's tiny. And what this is about is power, and not only in the electricity sense, it's about who decides. Does the electric utility industry get to decide whether their facilities are protected or not? Or, do they get to, as they are now, resist doing anything that anybody outside tells them to do, because it might save our civilization? Doesn't matter. They would lose power. They would lose the ability to decide whether or not that 20 cents per person was spent or not. And, from their point of view, that is a very bad thing. The 20 cents cost is not the issue. The issue is: can somebody tell them that they have to take a protective step? Or do they get to do whatever they want, even if they're putting the country and our civilization at absolutely huge risk? That's what's at issue.

CHUCK MANTO: Any of the other panelists want to chime in on that, or what local communities might do? And we have a couple over here. I'm going to walk the microphone to them. You can make a comment before I get there. Okay, we have two right here in succession. Identify yourself and then your question.

AUDIENCE MEMBER: Hi, it's Hank Cooper. And I think I know most of the people on the panel. I appreciate the fact that you talked about the threat of a nuclear EMP-generated attack. And I wonder if you could address the issue that I run into a lot, because that's an issue I care a great deal about, is defending against, in fact, with active defenses. People don't want to talk about that. And they think EMP is an excuse to build ballistic missile defenses. And that's why we're talking about this issue.

Generally, when I say what can be done about the threats you described, from missiles off our coast, and particularly the Gulf of Mexico, where we have absolutely no defense, I believe we have to have a hardening of the grid, because no defense is perfect. But we have to do the defense too. And the fact is that there are components of the EMP that we have to deal with, that come from nuclear weapons that we don't get there and get from some of these other threats that you were talking about.

So I wonder if you could address the psychological issue that we're running into in discussing this, what I consider to be the most comprehensive threat to the nation. And people don't want to talk about that, because they think it's, I don't know, too hard or too difficult or whatever, which is also nonsense.

CHRIS BECK: You made some important points, and I want to tie them into Ambassador Woolsey's points as well, such as when you talk about how a weather balloon would work just fine. So what we're faced with, as far as the nuclear EMP threat, is something that we really can't defend against. I think that there is some resistance in certain circles of looking at nuclear EMP because sometimes it's viewed as a stalking horse for reviving our own offensive nuclear capability, and things like that that people are shying away from. And I also have heard from the electric power sector, for example, in the United States, as Rick pointed out, there is a moving forward with the FERC order, but it's on geomagnetic disturbance only—it doesn't include EMP. The thought process there was, "Well, this is the electric power sector. They're not war fighters, they just provide power. So you know, nuclear energy is a national defense issue. And so that's beyond our scope—someone else has to deal with it."

The fact is: no one else can deal with it. You're absolutely right. Hardening, and being ready to take a punch and stand back up…there's no one else that can prepare us for that, except for the critical infrastructure owners and operators. They are on the front lines. And there are certain things we can do to lower the likelihood, of course, and there's no reason not to protect ourselves in any way we can. But pushing it off to say, "this isn't my problem" isn't a tenable solution.

CHUCK MANTO: Every question that is going to be asked will, of course, spawn five more, and lots of interesting discussion. And that's why we're all here, so you can get to know each other and continue this discussion. But, because we're going to try to keep things on track, I'm going to give the mic to one more questioner.

AUDIENCE MEMBER: Tom Popik, Foundation for Resilient Societies. There is a persistent rumor that Iran already has tactical nuclear weapons purchased on the black market after the fall of the Soviet Union, which would mean that they could have an EMP capability right now. In the opinion of the panelists, is this potential EMP capability of Iran affecting the Obama administration's negotiating strategy with Iran, its willingness to conduct a military move against Iran, and also the decision of President Obama in 2009 not to support the Green Movement, which would be a potential regime change?

PETER VINCENT PRY: I'll take that question because—thank you for throwing me the softball—because I wrote an article about that that got published last Friday. And the short answer is yes. But I would like to follow that up, if I could break your rules, Chuck, because I have a question. I'm seeing in this room, like in those three chairs over there, and Cindy Ayers and Claire Lopez over there, Frank

Gaffney in the back, and Sid Morris and Andrea Boland. It's amazing how much progress has been made by a handful of people, a handful of committed activists who have made tremendous sacrifices. Everybody on this panel, just a handful of people. And we've been trying and largely succeeding. Not fast enough, though, not fast enough, to get this country protected.

I have a question. My question is for you, you people who have contributed, you know who you are. The rest of you who are in this room, I know you're here because you at least care about it—you care about it or are curious about it. Please ask yourselves, what can I do? We need your help. We need your activism commitment. We need your brainpower. I'm not the smartest guy in this room. There are a lot of you out there who probably are a lot smarter than me, that can bring insights and talents to our cause.

I would ask you to go to our Uncle Sam. Frank Gaffney, Jake Berman, they've stood up an EMP coalition that's looking for volunteers. You know, if America is going to get protected, guys, it's not going to happen from NERC and the electric power industry. And unfortunately, Congress has tried and it can't do it. The President has tried and can't do it. It's up to you, just like in the old days. Rugged individualism and self-sufficiency are the kinds of values that established this republic in the first place, and it was expected to be sustained by the yeoman farmer and the small shopkeeper. I believe that that kind of spirit, that kind of American spirit still exists in this country. So that's my question.

JIM WOOLSEY: I don't know of anything plausible that suggests that the Iranians already have a nuclear weapon purchased from Pakistan or whatever. It's not impossible at all, but I don't know of any evidence to that effect. I do think people underestimate how far along both the Iranians and the North Koreans are to being able to launch an EMP attack. The North Koreans have nuclear weapons, and they are the right size—small yield, large gamma ray production—for an EMP attack. They have orbited satellites. So the nuclear capability and the satellite launch means that, as far as I'm concerned, they have the capability.

The Iranians have not yet, as far as we know, detonated or have a nuclear weapon. But they have orbited satellites. So, both North Koreans and the Iranians have orbited shooting to the south, as would be the case for an EMP weapon that uses the fractional orbital bombardment orbit that the Soviets basically invented, and we believe now have handed off to the North Koreans, and possibly Iranians.

So we are, whether the Iranians have a nuke yet or not, they are at most a few months away from being able to have one. And the possibility that they can have one very quickly, or already have one, and it could be an EMP weapon, is definitely plausible, as far as I'm concerned.

ROSCOE BARTLETT: I would like to note that the North Koreans have had tests where the rocket missile blew up above the atmosphere. We have said, "Gee, they really screwed up. And that's a failure, isn't it?" Jim, if you wanted to do an EMP attack, that seems to me to be a big success, isn't it?

JIM WOOLSEY: 200 miles out, that's about right.

ROSCOE BARTLETT: Chuck, I would just like to say, in closing, there are three levels we really ought to be working at in a matrix. The one I mentioned was, you know, your family and so forth, on up. The other is, we need to protect ourselves if we can. Our military is trying to do that. They cannot possibly be totally successful. So knowing that they cannot be totally successful, then we need to do the other two things. We need to harden the grid. And that's the primary thing we're focusing on today, what do you do there?

The third thing that we need to do is, what if those first two things don't happen? What if they can't protect us? And what if it happens? Then what? Shouldn't we have a pretty aggressive effort, at every level, so that we can survive? By the way, you're not going to be attacked where you're strong. You're going to be attacked where you're weak. Isn't this where we are potentially the weakest that we could possibly be in this area? Doesn't that invite attack?

Panel—The FBI, DHS, DoD and the National Governors Association Public/Private Plans to Mitigate Cyber Threats against Critical Infrastructure

http://youtu.be/WuaHqA5nUGw

Dr. Frank Kesterman, Professor of Cybersecurity and Homeland Security Management,
University of Maryland University College
Mr. Thomas MacLellan, Director, Homeland Security and Public Safety Division,
National Governors Association
Mr. Trent Teyema, FBI Washington Field Office, Assistant Special Agent in Charge
Mr. William O. (Bill) Waddell, Director, Mission Command and Cyberspace Group, Center for Strategic
Leadership and Development, U.S. Army War College
Mr. Robin Montana Williams, CWDP, Chief, National Cybersecurity Educa-
tion and Awareness Branch, Department of Homeland Security

FRANK KESTERMAN: The first speaker is Thomas MacLellan, Director of Homeland Security and Public Safety Division at the National Governors Association. The next speaker will be Bill Waddell, Director of Mission Command and Cyberspace Group, Center for the Strategic Leadership and Development, U.S. Army War College, Carlisle, Pennsylvania. Thank you for coming down here this morning. Next is Trent Teyema, FBI Washington Field Office, Assistant Special Agent in Charge of Cybersecurity, former Director of Cybersecurity Policy, White House National Security Staff. And then we'll have Montana Williams, Chief of National Cybersecurity Education and Awareness, responsible for the National Initiative for Cybersecurity Education. Our first speaker is Thomas.

THOMAS MACLELLAN: Great. Thanks, Frank. Well, good morning everybody. It's a pleasure to be here. It's tough to follow all the technical panels because I don't know anything in terms of the science of all this. But let me tell you what my area of focus is. I work with an organization called the National Governors Association (NGA). We are an organization that's comprised of the 55 governors out in the States, and I oversee what's called the Homeland Security and Public Safety Division. In my role, I get the opportunity to work directly with governors, as well as their Homeland Security advisors, some of their emergency managers, adjutant generals, criminal justice advisors, around a range of issues.

And I've got a short time here, but I want to provide to you an overview of a project that we launched about a year ago under the leadership of Governor Martin O'Malley from Maryland and Governor Rick Snyder from Michigan. It's called the Resource Center for State Cybersecurity.

The idea was that there is a lot of discussion taking place—and I'm going to tie it into some of the discussions we heard earlier, before—there's been a lot of discussion around cybersecurity taking place.

But it's all, frankly, been focused at the federal level. We have DoD protecting the dot-mil space, we have DHS protecting the dot-gov space. But there hasn't been a role yet carved out for states, for governors, what should they be doing to protect their own assets? What actions should they be taking to recover? And so with that in mind, we launched the Resource Center with that notion of filling that policy void for governors, for state actors, because from our perspective, they are an essential player to mitigating, to responding to, and protecting from any type of cyberattack—even if it's in the hands of a privately owned critical infrastructure.

Toward that end, we've put together a group of folks who have helped us think through what is it that governors can be doing? And governors are busy. They think high level, they think strategic. So we issued a call to action earlier this year, at an event on Capitol Hill—Governor Snyder came down from Michigan—and we call it a "Call to Action for Governors for Cybersecurity."

We recommend that states take five basic steps—we call it "Act and Adjust." It's the notion of "don't let perfect be the enemy of good." Because right now, the threat landscape is ever-evolving, it's ever-changing, and there are things that states can begin to do now, to effectuate as forward-leaning a cybersecurity posture as possible. That's the mission space we're in right now; we're looking at posture.

We will eventually move to looking at response issues, which takes us to the discussion here today. We don't care, at NGA, if it's an EMP (electromagnetic pulse), if it's solar weather, if it's seismic, if it's a cyberattack, with respect to the electrical grid coming down for any prolonged period of time. What we're going to begin to look at—and we are starting a process with an NGA that's going to probably align itself with our Cyber Resource Center—is this notion of response. How do governors (a) understand the primary, secondary, and tertiary impacts of a massive and prolonged failure of the electric grid? And (b) what are the responses? That mission space is also undefined.

Right now, one of the major concerns among Homeland Security advisors is this notion of catastrophic planning. What happens if something really big goes off in New York City? How do you evacuate Manhattan? How do you get the hospitals out? Part of that is what happens if the grid goes down?

We had a governors-only discussion back in summer of this year, and one of the discussions that we had was on this notion of the response to Sandy and other major catastrophes, tornadoes and so forth. One of the top concerns for governors is energy assurance. Without electricity, you have no comms after a certain amount of time. You have no fuel after a certain amount of time. You can't heat. You can't cool. So it's really been a major issue—and Sandy was big, but it isn't as big as some of the things that are possible out there.

One of the biggest threats, I think, that we face as a nation, with respect to cyber—in addition to the financial—is this notion of the electric grid being taken down or seriously impacted. So we're going to be looking at NGA to provide governors with advice, with resources. We're helping states right now improve their cybersecurity posture. Things called the "call to action." I'll just tell you the five actions that we're advocating for states.

One is this notion of creating a risk awareness culture. Two is this notion of understanding your own risk, your own threats, your own vulnerabilities. Three is really improving how you do continuous threat monitoring. Then there's also this notion of applying standard business practices. In particular, we point to the SANS Institute's Top 20 Critical Security Controls for Effective Cyber Defense—it's no longer SANS, it's now the Council on Cybersecurity. So there is a document. It's high-level, it's focused on governors. It seems light, but it's not. It is a strategic piece that we have put together for governors. It's really that notion of improving the posture.

In 2014, we are going to begin to look at response issues. I hope that helps set a little bit of the stage about where our thinking is at the NGA and really looking to expand the role of governors and states.

FRANK KESTERMAN: Thank you Thomas. Bill, please.

BILL WADDELL: Well, good morning. And I bring you greeting from the Army's best-kept secret, Carlisle, Pennsylvania, home of car shows and the Army War College. It's a great place to be from, and I always remind myself that working in Carlisle, the worst day in Carlisle is a 10 minute commute with one stoplight. So, when I come to D.C., it always gives me that perspective of where I work. So anyway, just glad to be here today.

When I was asked to look at the most important issue, or the top issue, that DoD faces in terms of its relationship with the private sector, I have to admit I spent some time trying to figure out which one would be the top one. But I came to the decision that I think that the biggest concern in DoD concerning this whole area of cyber protection and the private sector, has to do with the authorities and the relationships that go with this Department of Defense group of folks.

DoD is constrained very much by law as to what it can and can't do, in terms of reaction or how it works with private citizens. There are lots of folks just up the road in Fort Meade that have lots of capabilities; but unfortunately, they are very constrained in how they work with the private sector. There are two support missions that would work together with the Department of Justice, Department of Homeland Security, when these issues happen.

One of those would be the Defense Support of Civil Authority, or DSCA for those of you who are familiar. This is coordinated through the Department of Homeland Security, and it is in support of DHS's coordination themselves—being that DHS is the coordinator of all those issues. But some of the areas we've found in our series of war games, and some issues that we've discussed with the private sector is, there's still a belief that, if the whole town of Carlisle would come to Carlisle Barracks, that the cavalry would, indeed, ride over the hill and save them from this imminent problem that they face.

Of course, that's not true. We're not in a society of civil defense. We don't have rations; we don't have water. The DoD is dependent upon the private sector for its energy generation. It does have some generators, it does have some capabilities, but in a couple of days, they're going to be just as dependent as the private sector is.

Then you get into this area of authorizations, where again, DoD really cannot do anything—can't even share information with the private sector—unless they're authorized to do so. In many cases, that authorization has to come from the National Command Authority. You can imagine if suddenly, DoD was doing all of these things, the outcry from civilians, or just the government itself, would be massive. By nature of our Constitution, there has to be that control over DoD, but there are lots of constraints when these issue happen. So I see that as probably the biggest issue that I think DoD would be facing.

One of the other areas is that anything NSA does, and anything that U.S. Cyber Command does, of any significance, is very classified, and there are just not a whole lot of folks in the private sector that maintain those certifications, those security clearances, to be able to gather that information. So all of a sudden, you also have this information gap just because they can't share that information.

In case of a catastrophic-type issue like what we're talking about here today, this would be called a Homeland defense issue, where the national sovereignty of the United States would be in question. And once again, this would have to be authorized by the highest levels—the National Command Authority. And DoD would be part of the mission, but would not be directing the mission.

Now, for those of you who have done anything with the military, you know there are a couple of things that they always talk about—that's unity of command, unity of effort, those issues that go with that. Military folks are always talking about how that comes together. Unfortunately, in this scenario, what we've seen in a series of war games that we've done, is that none of the Big Three—Department of Justice, Department of Homeland Security, Department of Defense—is really established as the "lead agency." They each have their sectors, they have their responsibilities, but there is no chief belly button, if you would, who's responsible for all of those things. That can lead to those areas of real problems when you're talking about unity of effort.

So those issues would be Homeland defense mission, again, established by the highest level, the bringing upon of the capabilities of Department of Defense. But remember, there's that law that we talk about in all our classes, all the time, called "Posse Comitatus," that really is so restrictive of DoD doing anything against U.S. citizens. I agree it needs to be there. But we need to consider, and we need to come up with some policy on how we're going to do that in case of emergencies.

Which leads to my next point, which is, right now, due to a real paucity of policy at the highest levels, and that was discussed earlier, we don't have, in DoD, any standing rules of engagement. We don't have that list of, "If this happens, do this." That doesn't exist, so almost everything that DoD would do would be based upon either supporting Homeland Security, or would be waiting for some authority to come and say, "This is what we want you to do." That's a problem, and I think that that's where we really need to get to, as we move along this maturing of the cyber threat as a part of our Homeland Defense, Homeland Security.

Finally, just let me say that we have found in our war games that, from the civilian sector, there is a basic mistrust of calling the government in until—in several of our war games—until it's too late. When suddenly you're at a catastrophic place where the private industry has no longer anything they can do, then they call the government and say, "Fix it." That's a problem. We need to be working together.

And I will tell you that this organization, InfraGard and those ISACs out there, those Information Sharing and Analysis Centers that DHS coordinates with, those are the pieces, the places where relationships are going to be developed. That's where we're going to make the most money in putting this together, because without a policy, without issues, without specific missions established, you're going to have to work based on the relationships that are developed at organizations like this.

So I want to really commend the InfraGard organizations and those of you who are involved with ISACs, because that's how we're going to do this, is those personal relationships and crossing those barriers established by policy. Thank you.

FRANK KESTERMAN: Bill, we have a lot of work to do.

BILL WADDELL: We do.

FRANK KESTERMAN: Our next speaker is Trent Teyema, our distinguished FBI cyber leader in the National Capital Region.

TRENT TEYEMA: Good afternoon. My name is Trent Teyema, and I'm the ASAC or the Cyber Branch of the Washington Field Office. First of all, I'd like to welcome you to Washington, D.C. on behalf of the Assistant Director in Charge, Valerie Parlave. Welcome to Washington, D.C.—you're just in time for the rain and the storm coming this weekend, but welcome. Hopefully you'll get a chance to see our sites.

What I'm responsible for with the cyber branch, and I have bit of a long view, is if it has to do with a threat to the critical infrastructure, or has a spark, or any digital component, my group is the one that investigates it. So here in D.C. the way we're set up, you had Pete Trahon that welcomed you from our headquarters component earlier. I run the actual field investigations for the National Capital Region and Northern Virginia.

So basically, I had a colleague of mine who, when I ran the NCIJTF (National Cyber Investigative Joint Task Force)—usually we'd come out and do these threat briefings and tell you what the risk is— and actually he'd come out and say, "No, I'm here to do victim notification." Because usually, if you're in the National Capital Region inside the beltway, it's a major target center for cyberattacks from all kinds of different characters. And we're the ones investigating it.

With that, we work very closely, and usually—getting into what our role is, and kind of giving the perspective of what the FBI does, vis-à-vis DHS or DoD—we're the investigative component. We have 56 field offices across the country, 66 offices around the world. And so we're the investigative arm for the Justice Department. And we're closely tied between each office's cyber taskforce looking at cyberattacks against critical infrastructure, and then also tied closely with our Joint Terrorism Task Force.

What we do is we meld both a cyberattack and a physical attack against those critical infrastructures to investigate it. We work very closely with DHS and DoD, so when you have a call to one, you have a call to all. It's good to set the stage on what our roles are, because when we get involved, we're actually investigating, trying to find out what the threat is or the attack, and starting to mitigate it, and then we pass that information over to our partners at DHS, either working with FEMA, the NCIC (National Crime Information Center) or the different components, so that we can do resilience and protection.

Our investigation part also then feeds into what the Department of Defense needs and/or even the states and governors, so that they can protect from an attack. So, whether it's an individual group, a hacktivist, a state sponsor, a terrorist group, when you get down to the cyberattack, the tools and techniques are the same. It's usually the motivation of the actor [that varies].

So, when you're trying to exploit a system that's run by a computer control, you're basically trying to get that system to do what it wasn't designed to do. And then you're just trying to exploit it via code, physical, or even an EMP attack, to make it do something it wasn't designed to do. And we're going to try and figure out who is responsible and then trace it back.

Again, thank you for having us come out today, and I look forward to answering any of your questions. I think that's well under five minutes. And I'll turn it over to my colleague at DHS.

FRANK KESTERMAN: Yeah, we're making up good time here. Montana, please.

MONTANA WILLIAMS: Thank you. I'm Robin Montana Williams. I am the Branch Chief for our Cybersecurity Education and Awareness branch and lead the National Initiative for Cybersecurity Education (NICE), which focuses on an entire national program to do three things. The first is to raise awareness of cybersecurity from K through gray, as I say, of how to maintain good hygiene as an individual on the Internet, good practices at home, good practices in the workplace, and to understand that you're at risk in that environment and are exposed to a lot of different threats—and keeping that awareness up to be safe.

It's really no different than locking your home, wearing a seatbelt. You can look at a variety of different things that people do to mitigate risk.

The second piece to that is broadening the pool, which involves really enhancing and creating an environment where more and more people are entering the cybersecurity workforce. All my colleagues did a really good job about talking about the policy and the integration and how it's necessary to work together, and that, especially within the federal government, the three big gorillas, DoJ, DoD, and DHS, have very different responsibilities when it comes to cyberspace, and their roles in cyberspace, and working together.

But you cannot execute those responsibilities if you don't have the right people. A lot of people talk a lot about the software and the technical component and the hardware and those pieces of cybersecurity. But the essence and the center of most of our problem is the human part of it. I will ask this to the audience, and I ask most of my audiences: If there's any single person in this audience that can tell me that the APT, the Advanced Persistent Threat, used some sort of fancy laser and busted through all of our network systems and got in, come and tell me. Because that's not how it happened. It was a social engineering event; a human let them in the system. And that's where the problem is.

So you go from the awareness piece to broadening the pool, and then finally to the third piece: evolving the workforce. One of the things—and this was on the radio again this morning, people are still talking about it—is in the cybersecurity workforce across the country, there's really no defined pathway that takes you from what I refer to as "hire to retire." So, how do you grow a cybersecurity professional? Organizations don't have workforce development plans in place. What happens is poaching. One company goes and targets another individual from another company, offers them more money, offers them an opportunity to be promoted. And it becomes this cyclical event, of things happening where people just move from job to job to job. And you can't keep your best and brightest people, because you really don't have a workforce development plan.

That is the centerpiece of what we're trying to do in our office within DHS, and in the NICE initiative. DHS has many, many roles and missions in cybersecurity, from what was just described earlier of what the NCIC does, what the US-CERT (United States Computer Emergency Readiness Team) does from a technical standpoint, but also we see this as a bigger problem of actually helping and making sure that for the front lines out there, we're getting the right people, we're getting the right skill set needed to defend our networks.

One of the key aspects of this is the fact that if we don't raise the bar of our existing workforce, we're going to continue to fall further and further behind, and we're going to continue to lose this war. And I am sure many of you have read all the numbers, anywhere from $366 billion cost of known property loss, intellectual property, in criminal events alone.

So this is an epidemic problem. You've probably heard various reports. We need 300,000 workers in this career field. We need 1.7 million workers globally—additional workers in this cybersecurity career field. So the idea really, the center of what I'm really trying to do is to grow that. Organizations like InfraGard, the National Governors Association, we have tools and we have programs that we can work together, to help a state.

For example, the state of California right now is looking at expanding and really defining their workforce. They're using the National Cybersecurity Framework. How many people have heard of the National Cybersecurity Framework? Okay. So some of you who have heard of NIST (National Institute of Standards and Technology) have heard of the Cybersecurity Framework that way. For the National Cybersecurity Workforce Framework, which I want to introduce you guys to, what we've done is we've looked at the 31 key functional roles that occur in cybersecurity. The government is being mandated by OPM (the Office of Personnel Management) right now, and OMB (the Office of Management and Budget, in the White House)—to start building and defining their workforce based on the National Cybersecurity Workforce Framework. Thirty-one functional roles with the associated knowledge, skills, and abilities needed at those levels. That exists out there.

We have tools and things that are sitting on what is called the NICCS portal, the National Initiative for Cybersecurity Careers and Studies portal, which is a resource that is available to anybody from a parent to a governor to a CEO to a human resources person to a CISO (Chief Information Security Officer) on tools that they can use to help develop and train their workforce based on the Workforce Framework itself.

There are things that are going on, right now, from the critical infrastructure framework that NIST is working on, that came out of the EO (Executive Order), all the way to the Workforce Framework, which is helping feed the opportunities for folks who are in this career field.

So I'll stop there, and we can start engaging and answering questions. But what I want to reiterate is the fact that the workforce and the human aspect is just as important as policy, just as important as the hardware and software, and it is an issue that we need to continue addressing, and that sometimes I think is neglected by people who are thinking in this lane of the road. Thank you.

FRANK KESTERMAN: A lot to think about. Panelists, do you have any questions of other panelists? Do you want to add anything from what you've heard?

BILL WADDELL: Yeah, I want to address my question to Tom, and that is, one of the areas that we as a War College are moving toward is the development of how the National Guard Title 32 forces integrate into the cyber workforce, perhaps from a Title 10 DoD perspective, but maybe more significantly how that would affect the governors and the governors' capability of doing that. Have you guys done any kind of research or work in that area?

THOMAS MACLELLAN: Yep. A number of years ago—it was actually set up under President Bush—what's called the Council of Governors is 10 governors appointed by the President of the United States. It was ostensibly designed to focus on DSCA and to look at unity of effort, unity of command issues, with respect to DoD, DHS, so those principals are at the table.

A couple of weeks ago, I was at one of the staff sessions of those with DoD, with DHS—I think DoJ at some point will be brought into it—and right now, we're looking specifically at defining the mission space for cyber between states, DoD, DHS and so forth. In fact, we're actually working toward an agreement, an MOU (memorandum of understanding) or some vehicle similar to what we did for DSCA. That is really going to help define the mission space. It's a starting point. It's going to help begin the discussion of that, and it's going to help define that.

It is a big and very complicated issue. It brings in Title 10, Title 32, the role of the National Guard, budget issues. It's a very complicated negotiation and discussion that, when the governors come together in February this year—we bring the governors together twice a year—when the council meets, that will be topic number one that we're going to be working on. So if I could throw out two responses—is that fine, Frank?

FRANK KESTERMAN: Sure.

THOMAS MACLELLAN: So two responses—one of the things I didn't mention is that we at NGA, you know, one of the focuses that we are looking at is that notion of developing a skilled workforce, and what tools governors have available to them. So I want to just make sure that you're aware that we're working, we're aware of your tools, the NICE. And we're actually pushing that framework out to states.

But the question I want to ask is to our FBI guys. And one of the things that we are also emphasizing to governors is expanding the role of fusion centers in the cyber world. There are a few fusion centers that are out there, that are beginning to stand up some capabilities. I'd be interested to hear your take on what you see their role to be with respect to fusion centers.

For those of you who don't know, there are 78 fusion centers out there. They were originally designed to collect information around terrorism. The mission has involved—there's been a maturation process, if you will, with respect to some baseline. And they're beginning to move into cyberspace. I'd be interested to hear your take on that.

TRENT TEYEMA: Well, with the fusion centers, I think what we need to do is they need to evolve, and they have begun evolving. It's not just a straight terrorism or counterterrorism mission, but actually getting into that information sharing and actual response, because what we've seen, both with our Joint Terrorism Task Forces, our fusion centers, and then also cyber taskforces, are our membership from state, local, tribal as a part of that, and through that, we're able to share.

In some parts of the country, the fusion centers are like models and working amazingly well. Others need to mature. Right now, from what I've seen in my experience, it depends on the area and the involvement, by the state and locals, how successful the fusion center is. Some have their framework, and they're passing information. Other fusion centers are just putting points on the board right and left, and it really is doing what it's supposed to: you're getting all that information into one place, and then being able to share it.

That's always the trick, sharing that information. Because coming down to it, 90-some percent of the information is really open source or unclassified, maybe law enforcement sensitive, but a very small percentage is actually classified, which—that ends up slowing everything down. We've seen it over the last two years, as the government has gotten away from getting tied up in the classified or over-classified information, sharing the information that's just out there. It's in that speed of information and building that speed of trust that you have the success, I think, if that answers your question.

MONTANA WILLIAMS: Just to add a little bit to that, the MS-ISACs (Multi-State Information Sharing and Analysis Centers), and stuff that we've worked with our partnership with the FBI [have found that] Trent's correct—there are certain fusion centers that are just awesome, and some that just need a little bit of work. So one of the things DHS is really trying to do is gather up best practices among the fusion centers, share them, work on a common taxonomy. Because sometimes there's this lovely language differential of terms that different people use. And so, when you're talking to somebody at the US-CERT (U.S. Computer Emergency Readiness Team), or you're talking to somebody at the Joint Task Force for the FBI, getting the right dialogue so everybody is kind of on the same page—looking at ways to train everybody to some basic level within the fusion centers, so they're getting some of the training that some of the government folks are getting—those are some of the things, as we move forward, over the next few years, that we're really trying to push from DHS's perspective also.

THOMAS MACLELLAN: But on that notion, I mean, isn't one of the biggest challenges to some of the information sharing—take fusion centers—is that the cyber mission space still isn't well defined. I mean it's defined there, but it's defined within a law enforcement culture. So you have the CIOs (Chief Information Officers) and the CISOs in the states, who aren't law enforcement, and often can't get access. And then you also have the issue with private sector entities who may not trust, who may not want to share.

How do we begin to overcome some of those things, where the cyberspace is—it is a law enforcement issue, it's a critical infrastructure issue at times. And you've got the CIOs, the chief information officers, or the CISOs, who have this enterprise-wide view, but may not be in the law enforcement culture. And there are barriers that they need to break down. I'd be interested to hear any of the panelists' reactions to that.

TRENT TEYEMA: That's exactly true. The most success we've had is through InfraGard. I mean a lot of this is building that speed of trust and actually bringing them in. I meet with C-suite level individuals all the time, CIOs, and we want to get them to our InfraGard coordinators, get them on a first-name basis, so when they have an issue, they can deal with it directly. I spend most of my week getting hypothetical calls from CIOs and CISOs, saying they have a problem. [It's important to bring these people together], building that speed of trust—I keep going back to it—so that, you know, there is that myth that you call law enforcement, and then you have the FBI show up, there's a whole bunch of gray jackets. Same thing with DHS; they're going to show up. You have a whole bunch of government types that are going to slow down the business. And that's not what we're here to do.

A lot of the conversations I have is, "Hey, I heard something's going on. What are you hearing?" And it's literally both ways, because though questions like, "Is this a real issue?" "Is there a new DDoS (distributed denial-of-service) attack going on?" "Is there a new zero day exploit going on?" may not be directly to their infrastructure or their enterprise, you're having that dialogue, and it's growing.

I started at our Los Angeles chapter of InfraGard 15 years ago, and have been involved in it on and off during my career. Our Los Angeles fusion center, you know, like Baltimore's and D.C.'s chapters, I mean really, it's a model for the nation. I'm always so impressed at how involved our membership is, as opposed to other chapters, where they're just meeting once a month and, you know, are not really as engaged.

So, you know, your point is exactly right. Depending on if you call the D.C. office, or you call another office, you may not get the same response. DHS—and Montana, you could probably go into this—is more centrally located here in D.C. They have the PSAs (Protective Security Advisors) across the nation. Again, it depends on the outreach. We try and give them the same response, but you have to build this culture and bring it in. It's an excellent thing to highlight.

MONTANA WILLIAMS: Yeah. It's relationship-building at its core. So we have our PSAs, and now we're starting to infuse CSAs (Cyber Security Advisors), who are our cybersecurity folks out there.

FRANK KESTERMAN: Do you want to explain PSAs?

AUDIENCE MEMBER: Public service announcement, right?

MONTANA WILLIAMS: No, not quite. So our PSAs work with critical infrastructure. And those are agents out in the field. They're assigned, right now, to the FEMA regions. It's built on the FEMA construct. So there are PSAs located all over the country, various large cities that have significant activities going on there that really affect critical infrastructure. You can get on the DHS website and find out points of contact for your PSAs.

The cyber folks, we just have four or five of them right now, distributed across the country—they're overwhelmed. We're trying to get more billets to put more folks out there. I'm really working hard for the Las Vegas job. I'm publicly announcing that right now.

The job for them is exactly that: it's working with all the other government agencies and linking with the private sector, working with academia. It expands across the entire community. And not just including the 16 or so, 17 critical infrastructures, to basically help solve problems, provide points of contact for people to reach out to.

And, as Trent was just saying, it's got to be relationship-building, where there's a trust that's established, where companies are comfortable coming to somebody from InfraGard or somebody from a specific agency, and sharing information, and knowing that it's protected. Remember, when you're dealing with the private sector, it's all about stockholders, and it's all about their reputation. For a private sector organization, their reputation and their brand are actually more important than anything.

They tend to keep a close hold on a lot of that stuff, but we've found that over time, as different events have happened, that there's a benefit of this information sharing among the various companies. A lot of companies are coming around and wanting to participate more, but that's driven a lot, in a lot of cases, and even with most states, by the personality of that PSA or that CSA or that InfraGard chapter out there that's creating this involvement.

It really comes down to adopting these good ideas, adopting good hygiene practice, adopting good information-sharing practices. It all works together, and it's very, unfortunately—fortunately or unfortunately, because cyber, we don't tend to—a lot of geeks don't like to be outgoing and engage with folks. But that has to happen. So you have to have people out there who are engaging, building and developing relationships.

FRANK KESTERMAN: Great comments, Montana. I want to shift to Trent to ask him about information sharing between the FBI and membership in InfraGard and others.

TRENT TEYEMA: Okay. If you're not a member of InfraGard, I suggest that you join. It's free. That's my free pitch here. It doesn't cost anything. The whole purpose of InfraGard when it started was infrastructure protection, and cyber is a big component of that.

InfraGard is also the mechanism the FBI specifically uses to push threat analysis and tippers, cueing—the whole thing—out to the community. That's really grown over time. We closely partner with DHS over there. So like, when we're investigating something, and I discover a new threat, a new zero day, a new compromise, a new attack, we're going to push down through our InfraGard network first. When you join

InfraGard—some background, if you're not a member—we do a light background on you. It's a closed network, so you can share information behind the fence line. But we can then quickly push information out.

So when I get threat information, not only do I push it out through the InfraGard network, I'll push it over to DHS, and then have them push it out through their network. I like doing that because, if I have an investigation going on, I don't want to tip that the FBI is investigating this, so I push it out through the normal networks. And DHS is amazing. Their reach, as they go out, they're pushing it out through the normal feed, which I recommend. You know, it's like Patch Tuesday. Pay attention to that. Because a lot of the investigations that we're working, and the other investigative agencies, we utilize DHS to get that out. And that's how you're going to be able to know that the threat is out there.

A little bit on protection: through InfraGard we have a reporting portal called iGuardian. iGuardian is a network through which you can report threat information securely to us. It comes in, it's immediately actioned out of headquarters, and then assigned to the field office so we can start going on these threats. So it's everything that's not just a cyber reporting. It could be, you know, somebody is acting suspicious. It can be everything from criminal to terrorism to counterintelligence to cyber, across all those.

iGuardian is that portal through which you can enter that information and we can respond. And then basically, as I mentioned before, a call to one is a call to all. That's one of those intake points that usually we get: Who do you call? Do you call Montana? Do you call me? Do you call the Governors' Association? Once we get it, we try to share it between all the organizations so that you have awareness.

And then it's usually the back office calls: "Okay, are you looking into this?" "Yeah, I'll get back to you." Or, "What do you know about it? Is this a real threat, as I mentioned? Or is this just one of the thousand DDoS attacks that are going on every day?" And it's one of the reasons to be involved in InfraGard, because it's a way for us, as an organization, to push information out to the public.

FRANK KESTERMAN: Okay. Time to go to the audience. Yes, sir. Please identify yourself.

AUDIENCE MEMBER: I'm George Baker. I'm a professor emeritus, James Madison University. EMP and RF (radio frequency) weapons have been called the ultimate denial of service attacks. And I was wondering, to what extent do you include those effects in your planning and in your programs? I'd be interested to hear from any or all of you.

TRENT TEYEMA: Well, I can say what we do at the FBI. So we work very closely with DHS, ICS (Incident Command System), CERT. So we have agents that we train specifically for that, we send them through Idaho National Labs. They're usually part of our regional deployment teams. When you say cyber to people, you never know what that means. Depending on what the rubric going into it is, you always have that filter.

We investigate threats against something that has a spark usually, something that's going. So EMP and RF weapons are real threats. So we team up with our domestic terrorism teams and our IT teams and kind of look at it as a holistic team. When there's an issue, usually a call goes to both us and DHS, or we call each other, and then you have a blended team that's going out to it.

Have you guys seen Make Magazine? It's on the newsstands. Make Magazine comes out monthly, it's an engineering magazine, so you can make robots and all kinds of interesting things with electronics. This month's issue actually has stuff on making your own personal EMP device, and how to do it, and gives you the schematics.

So, as you're talking about it, it's not that hard. But it's out there on the net or on the newsstand on how to do it. So the range of characters that we investigate are everything from fringe to nation state. It's definitely in our training and what to do—because an EMP or an RF pulse against electronics is usually very bad—and then how to respond to it.

MONTANA WILLIAMS: That's exactly how it works.

BILL WADDELL: The bad news from a DoD side is, although they discuss EMP, they don't really talk about what they're going to do when it happens—at least at the War College level. We had a war game a couple years ago that discussed all of those areas, and the EMP scenario was basically, "Beam me up Scotty. I have no capability." It's not well covered at the War College level. There are so many other issues that go on, I'm just glad we can at least get some cyber into the War College curriculums. But that's definitely a future topic, as somebody at a level well above me identifies that this really is a problem.

MONTANA WILLIAMS: Yeah, and I'm an ex-B-52 aviator. So, and the DoD was all about the delivery of EMP.

CHUCK MANTO: We have another question from the audience, unless somebody else was going to answer that other question.

FRANK KESTERMAN: Bill, I want to ask you about power plants on bases. Is there a program that's moving along on that? Like, you wanted to reduce reliance on the private sector power plants by building utilities. Do you know anything about that?

BILL WADDELL: I will tell you that I know that the Corps of Engineers and some of your engineering capabilities across DoD are aware of the fact that, when this happens, they will generators until they run out of fuel. But I will tell you that there is no real program that says we're going to create our own energy-generating capabilities so that we don't have to rely on the private sector. I don't think that program is out there. I could be wrong, but I don't think so.

MONTANA WILLIAMS: It's not cost-effective.

CHUCK MANTO: Okay, we have one final question in the back here.

AUDIENCE MEMBER: All right. I'm a student at UMUC. I'm curious about the human aspect. You might have the best firewalls, the best security system. What about internal human aspects in the government that could create leaks on cybersecurity issues? What are you doing about that?

MONTANA WILLIAMS: Well, we read all about it in the press. And, if you ever spent any time with Kevin Mitnick who coined the term "social engineering," and if you don't know who Kevin Mitnick was, he's actually the first person who was ever prosecuted and spent time in jail under Internet crime law. He is one of my neighbors down the street in Las Vegas, actually. And he will tell you that you could spend a billion dollars on cybersecurity, it just takes one stupid human trick. So, we'll put David Letterman in the whole phrase here too, "Stupid human tricks."

The essence of that is that insider threat is one of the greatest fears to most government or our DIB (defense industrial base)-type organizations, or even your Targets or your Sears and Roebucks. It's somebody inside who has access and can do undue harm. Part of that is training and awareness for your entire workforce on how to recognize the insider threat, and how to deter it.

I wrote a paper a while back on the idea of what happens in workplaces. There are various things that occur in workplaces with cyberspace. One was cyber slouching, which means you just sit on the Internet all day, and you don't really work. And that never happens in any organization I've ever been involved with.

But part of that core essence is, when you're looking at the human being, what are their motives? Some of the organizations do a little bit of continuous monitoring within their organization of what people's activities are; they block things. There are things that you can do from a policy standpoint.

But you have to deter those types of actions. Almost 80 percent of all civilian, or private sector, companies do not have any type of Internet usage policy in place in their organization. And it may even be higher than that. I'm probably being very conservative when I say 80 percent—it might be closer to 90 percent do not have a policy.

So if you're doing something bad on the company's network, how are you published? What are the repercussions, other than a slap on the wrist, or a little bit of a scolding, or anything like that? So, if you're going to drive human behavior, remember this. It involves consequences. There has to be a consequence for action. You know, I hate to say this, but we're all kind of still children when it comes to this. And so, unless you get your hand slapped once in a while, or it hurts you—you only have to stick your hand on the burner one time, and you'll probably remember you're not going to do that again.

So it really does boil down to a policy set in place by organizations, whether they're government organizations or private sector organizations, that there are repercussions for negligent activity or malfeasance on the Internet. That's the bottom line.

BILL WADDELL: Just one quick comment if I could. One of the smartest things DoD did in the last five years concerning this issue is that they are now holding commanders responsible for the things that go on in their commands. This is not just a "Sparky you got it," or what people would know as a six, the communicators folks. It's the command responsibility.

THOMAS MACLELLAN: Just real quickly, from the governors' perspective, we're pushing this notion of creating a culture of awareness. You can see it manifested in two states in particular, Maryland and Michigan. In Michigan, Governor Snyder has pushed out a brand new training package for all state workers. It's an online thing. He takes it, he asks his cabinet members, "Hey, have you taken this various pod?" Apparently, it's effective. Maryland has taken the same approach. And they're also looking at ways to actually make that part of the review process, not just the taking of the course, but other kinds of things as part of like an OPM-type policy.

MONTANA WILLIAMS: That's a benchmark. We're going to take the Michigan program and push it out, as hard as I can, nationally. That is an incredible program. But that's from a governor who has a cyber technical background. So he has a love for this.

Latest Idaho National Laboratory Research Data on GMD Impacts to Power Grid Infrastructure

http://youtu.be/Ynpi7zW-SL4

Mr. Scott McBride, National and Homeland Security, Critical Infrastructure Protection & Resilience, Idaho National Laboratory

MODERATOR (CHUCK MANTO): It's my pleasure to introduce Scott McBride, who is going to tell more about his role at the Idaho National Laboratory and their work testing a live power grid for vulnerabilities to solar storms.

SCOTT MCBRIDE: Thank you, Chuck. I'm happy to be here to talk to you today. The Idaho National Laboratory has been working on a program with the Defense Threat Reduction Agency (DTRA) and Scientific Applications Research Agency out of Colorado Springs for the last three years though that program is older than three years. It started when DTRA built a test facility in Albuquerque, and they did some initial ground induced current testing on a small distribution system with a couple of small, distribution transformers.

During the course of that testing, they were introducing direct current (DC) into these transformers, driving them into half-cycle saturation and then looking at the harmonics and losses that were generated from that anomaly. So that test basically simulates the effects of a geomagnetic disturbance on a distribution grid. They realized that in order to bring that closer to reality, they needed to step up and test on an energized transmission grid. At the time they were using a small, diesel generator as the source to provide the power to that test grid.

So they stumbled into Idaho National Laboratory. We are a Department of Energy Lab, and we have a unique infrastructure in that we have 890 square miles of federal reservation. It's about the size of Rhode Island. And we operate our own utility grid. We have 61 miles of 138 kV transmission. So I'm going to show you this slide deck that is mostly photographs. I'm going to try to keep this to the fundamentals and the basics. And if you have interest in more technical reports we will have those available here down the road.

So the first thing, in order for us to be able to test on our grid, which has seven main substations, including several operational nuclear facilities, we operate the advanced test reactor. We operate the materials and fuel complex. So we develop new, nuclear fuels to remove high-enriched uranium out of research reactors in the 'stans (Kazakhstan, etc.) as we call them, in order to replace that with low enriched uranium, which is less likely to be made into a weapon. So those facilities are operational class 1E nuclear facilities and we can't disrupt the operation of those facilities. And, yet, we are going to start introducing these anomalies of ground-induced currents on our transmission grid.

The photograph on your left shows what we call the Real Time Digital Simulator (RTDS). That is a system that was developed in Manitoba Hydro. Manitoba Hydro spun out and they formed this company, RTDS. It's now used worldwide: China, India, they use it for doing dynamic simulation of new transmission and generation facilities. At Idaho we use that—we develop a physics-based model, that operates in real time, of our power grid. We introduce the anomalies that we are going to do during our testing, and then we are able to simulate the effects that we are going to see on our transmission grid in order to prove to our operational nuclear mission owners that we're not going to subject them and cause them to trip offline.

So first we model, then we come out to our test site. Now in that diagram, the outlying, odd-shaped block shows our 890 square mile reservation. We've built 52 test reactors on that facility starting in the 1950s. The site was actually commissioned as the NTS (National Technical Systems) and we tested the Army guns after World War II. They were shipped south of INL about 50 miles. Those barrels were relined off the battleships. They were shipped by rail to Idaho National Laboratory and then we test fired those guns and boresighted them. So that's why we have so much real estate.

So the inside, that loop shows our 61 mile transmission grid. At the bottom, which you probably can't see, we have an area called CITRC, with is Critical Infrastructure Test Range Complex. And to the right is the MFCTS, or test site, which is right adjacent to the Materials and Fuels Complex. In between CITRC and MFC Test Site we have 13 miles of 138 kV transmission. We are able to open that transmission loop and leverage that transmission line for these tests. So at the CITRC location we have loads, and at the MFC Test Site I'm going to show you, we've built a temporary substation and that's where we actually do these tests.

So this is the MFC Test Site. And I'm going to point to a couple of things. First of all, I'm going to point to the two power transformers under test. And then I'm going to point to our test trailer or load trailer. These are the transformers. This is the load trailer. So the two transformers I pointed to, those are 138,000 volts on the primary side, with grounded wye winding for you electrical engineers. And the secondary side of those transformers, one of them is 13,800 volts. The other one is 2,400 volts. Those two transformers were installed in INL substations back in 1954 and 1955.

They operated there for over 50 years and then, just based on measuring the oil, dissolved gases and moisture in the oil and insulation power factor tests that we performed, we determined that they were at the end of their suitable design life. And so we purchased new units and replaced those. But instead of salvaging or throwing them away, we put them in a bone yard. Well, in comes DTRA and wants to do these tests with power transformers. So we hired a crane and a transport. We moved those transformers into location. We built a temporary substation, including all the ground grid and oil containment ponds and other features that we needed. We built that right under the 138 kV transmission lines that you see overhead.

Now on the transmission line you will see some bags. I'm going to point to one of them now. Those bags are associated with current transducers that we have on the transmission line. So we're injecting currents into the neutrals on the transformer, and then those currents propagate through the transformer windings and they end up on the transmission lines. So we're monitoring voltages on 138 kV, currents at 138 kV, as well as frequency and all the harmonics that are generated. We are also monitoring voltages and currents and frequencies on the low voltage side of those power transformers. And then inside the load trailer, which I'll show you a photograph of, we're monitoring voltages at 480 volts all the way down to 12 volts.

So also we have in the bully barn, in the bottom right, that's a UPS, uninterruptible power system. To the left of that is a 30 ton chiller. And then in the very back left are three resistive load banks. So we're able to conduct these tests with both resistive loads, which would be like giant space heaters, as well as complex loads. So an industrial or commercial facility has a lot of complex loads. Those include variable frequency drives, pumps, fans, uninterruptible power systems. We have a whole bunch, like maybe a couple hundred computer switching power supplies inside the load trailer.

I think SARA, Inc. (Scientific Applications and Research Associates) actually assembled that trailer for DTRA. And that trailer is designed to replicate a critical DoD facility. Now there's a couple of distinctions. DTRA is interested in really providing the knowledge and information to protect critical DoD assets from geomagnetic disturbances. Industry, or the electric utility industry, they're interested in knowing what they need to do to protect their infrastructure or whether or not geomagnetic storms are a potential threat.

So doing this test for DTRA, sponsored by DoD, we're able to leverage the results from these experiments to also potentially assist industry and the government in developing effective policy.

The office trailer that you see there, the 24 foot office trailer—all of the data acquisition signals that are collected through the data acquisition systems that we have distributed around the test site and at CITRC substation, all of those signals come into that load trailer, which is where Amber Walker (from SARA, Inc.) sits, who is in the back there. And she actually executes the tests. But all those terabytes of data are then recorded there so we can go back and review that data after the fact.

This shows the inside of the load trailer. We have CPU loads, which are the computer-switching power supplies I mentioned. We have a Chloride UPS and an Eaton UPS. On the very left-hand side we have a HEMP (high-altitude electromagnetic pulse) filter. Anybody familiar with a HEMP filter? A few of you are. That filter is installed at locations like data centers, call centers, critical DOD mission facilities, hospitals, potentially, and other critical loads. That filter is designed to protect against the E1 and E2 components of the HEMP pulse. E1 is a very high magnitude, very short duration pulse. E2 is from like one nanosecond to one second.

E2 is similar to lightning, is what we often refer to. We know how to protect against lightning and we know how to protect against E1, by shielding electronics and hardening. E3 is the low (frequency) and slow component of the EMP pulse, which is what is significant with geomagnetic disturbance potential impacts on the power grid. So the tests we're doing are to simulate the E3 component of the HEMP pulse. Well, that filter on the end of that load trailer, we were able to do these tests and measure the voltages and current and harmonics at the end use loads in that trailer with that HEMP filter in the circuit and with that HEMP filter bypassed.

A key takeaway that we found: when we do these injection tests, the harmonics we measured inside the load trailer were about 50 percent higher with the EMP filter in the circuit than they were without that filter in the circuit. So now we've got a filter designed to protect against one challenge that exacerbates the problem due to geomagnetic disturbances.

The first year we tested, we weren't sure exactly how this was going to play out. So we injected up to 128 amps of DC current into transformers that were loaded at approximately 10 percent. Well, most transformers that are out there are loaded at 50–80 percent, maybe even 90–95 percent. So we had light load on the transformers. The second thing, we wanted to lower the fault current that was available at our test site. At the CITRC substation 13 miles away, we introduced an artificial impedance by putting two power transformers in series to limit the fault current at our test site and also to basically choke any harmonics that we were going to send back to the rest of the INL operational facilities.

There was one other change we did in the last year of testing over the previous year and that was we tested a mitigation strategy, which I'll talk about a little bit. And then you'll hear much more about that this afternoon.

So the device on the left is the Emprimus neutral ground blocking device, which has been licensed to ABB. ABB is marketing it as "Solid Ground." That device is designed such that if its control system detects DC currents on transformer neutrals, it engages a series of AC and DC breakers. And it shunts those DC currents to ground and allows the substation transformers to ride through the ground-induced current anomaly. So think hospitals, law enforcement, critical DoD missions, critical substation locations. This is a hardware device that we tested to mitigate ground-induced currents and allow that substation transformer to ride through.

On the right are Schweitzer relays. We had a bunch of Schweitzer engineering folks at our laboratory. They brought a bunch of their relays. They collected a bunch of data while we were testing. And Schweitzer is looking at developing algorithms that could be upgraded into firmware of existing relays, which are worldwide, that could potentially detect these same DC currents on the transformer neutrals. But instead of allowing the transformer to ride through, it would open the transformer differential and de-energize the transformer. Once the anomaly goes away, that transformer could be re-energized. So you're protecting the transformer but the customers will see an extended outage.

A couple of key points: the second year of testing, we removed that artificial impedance that we introduced in the first year of testing so that we had a much more realistic test, with full fault current availability at 138 kV. We also loaded the smaller transformer, which is roughly 3.8 MVA, or megavolt amps. We loaded that transformer to 100 percent. And the larger transformer, which was 15 megawatts, we loaded that transformer to 50 percent, both with complex and with resistive loads.

Our loading was much more realistic, our transmission source was much more realistic, our results were much more realistic. And I believe the industry would accept the results of the more realistic test much more readily than the first test that we did.

One of the things that we found out is that the stiffer the source, or the lower impedance of the source, the more harmonics the system can absorb without causing problems. So the first measurements that we made in year one [showed that] the harmonics were much greater than they were when we tested in a much more realistic way. These transformers that we tested are all of the same design. They are core form, wye-grounded primary, three-legged core transformers. There are a lot of comments amongst the transformer designers as to which transformer designs are the most vulnerable.

We have shell form transformers, core form transformers, and autotransformers. High voltage transformers, 345, 500, and 765 kV—a lot of those transformers are autotransformers. We have not yet tested an autotransformer. That's one of our next steps. We're searching for an autotransformer right now. We would like to test an autotransformer similar to what we have tested the core form transformers. And then we would also like to heavily instrument a transformer on the core and in the windings for temperature and vibration and be able to test a transformer to destruction.

Power equipment runs on smoke. When you let the smoke out of it, it no longer works.

Are there any questions I can answer?

AUDIENCE MEMBER: I'm Bron Cikotas. The question I have for you, that filter you showed, it will work for E1 and E2. It will not work for E3 because it's essentially a quasi-DC signal. So maybe a clarification is needed there. The other question I have for you, have you tested delta-wye transformers, where you can essentially break the incoming ground and pick up a ground at the site?

SCOTT MCBRIDE: First of all, your first question, the E1, E2, E3 on the filter, yeah, I do understand the filter is designed to protect against E1 and E2 and not E3. What we found, though, was that the E1, E2 designed protection filter increases the damaging harmonics that would cause impacts to the end-use loads when you give it an E3 signal. So it exacerbates the harmonics that we were able to empirically measure.

BRON CIKOTAS: The timing of those signals are separately, sufficiently apart so you won't have that interference in a real situation.

SCOTT MCBRIDE: Well, the pulse that we're injecting is eight seconds. So we have a one-second pre-trigger, an eight-second DC pulse and then a one-second post-trigger. And then we record the data during that 10 second period. The second thing, the transformers we tested are both grounded wye high sides with delta low sides. And then we have a zigzag transformer and a grounding resistor to derive a neutral on the delta windings on the low side. That's not all that common in the utility industry. Most extra-high-voltage power transformers would be delta on the high side with grounded wyes on the low side.

Updates on Space Weather Threats
for Power and Communications

http://youtu.be/k_XICPoZLeY

Mr. William Murtagh, Program Coordinator, National Oceanic and Atmospheric Administration/National Weather Service Space Weather Prediction Center

MODERATOR (CHUCK MANTO): Your presenter, Mr. William Murtagh, Bill, is a top expert from NOAA's Space Weather Prediction Center, which is responsible for operationalizing space weather observations and warnings. Right now I'd like to present Bill Murtagh.

BILL MURTAGH: All right, Chuck. Thanks so much.

It's unfortunate that John [Kappenman] is not here. Our presentations are usually very complementary because I'll present the science and the space weather piece and John hits on the "so what." But I'll try to cover a little bit of both in the presentation here today.

Now, one thing I'd like to clarify for a lot of people in the audience who are seeing stuff for the first time, when you look at slide you will see we are part of the National Weather Service. And people in this country tend to say, "What is space weather doing in the National Weather Service? Why is it not in NASA?" And I'd like to point that out just because it is a reflection of where this issue of space weather has come today. Because, indeed we were more of a research and geeky type of group back 10, 15 years ago.

But as we became more and more dependent on advanced technologies for everything we do today, we recognized that space weather was a real threat, a real and current threat on that very technology that we rely on. So no longer did we belong in any kind of a research-y agency, we belonged in an organization that issues alerts and warnings. And that's what we do: we issue the alerts and warnings like any weather station in the National Weather Service…except I don't really care about the tornadoes and the hurricanes, the blizzards and whatnot. My interests lie way above that—on the solar flares, geomagnetic storms, and radiation storms and how it can affect that technology.

So in the Space Weather Forecast Center—and you are very welcome to come out to Boulder, if anyone is ever interested in coming out. Don't come on a day like today or yesterday where it was 15 below zero when I left Boulder yesterday morning. But we have an operation center that is staffed 24/7, providing alerts and warnings for the nation and, indeed, our colleagues around the world.

What I wanted to hit on today is a little bit about the solar cycle business. Sometimes it is somewhat misleading and misunderstood. And close calls, a couple of key messages—I want to make sure when you walk out of here today you understand some of these key messages, so I'll highlight them and some of the impacts—just some of the more rubber-meets-the-road type of stuff, because I know there are folks out there that hear about this space weather stuff and they just don't quite get why it applies to them and whatnot. I hope I can leave you with some kind of understanding why one should care about it.

This chart here, and I know for the folks in the back it's busy—it's supposed to be busy. It's the reflection of what happened back in October 10 years ago, October 2003. We refer to them as the Halloween storms. This was the wakeup call. It was at this point in time that we really realized—we've had some activity in the past that did cause problems—but the widespread and diverse impacts at this period of time in late October was a wakeup call. We recognized, hey, when we don't have GPS, that's not a good thing. If we don't have GPS for a few minutes, it's not a good thing for some people, but if we should lose it for days upon end or longer, it would be very, very significant. Any kind of satellite communications [would be degraded] and, of course, as everyone in this audience is aware, we have concerns with our electric power grid.

But you can just see across the world—and it's one of the takeaways from this slide—it's recognized, probably like no other natural phenomena: these big, big space weather storms are global. They are not going to focus in on one area. There will be localized effects and whatnot but we could be seeing power grid problems in Canada, in the northeast United States, and at the same time, the Ascom network in South Africa could be having significant problems. It is a global phenomenon.

I wanted to talk a little bit about the solar cycle and I'll key in on the takeaways here. When you look at this plot, you'll see that graph, the blue line back in the horizontal scale on the bottom is the year. We had our last solar maximum in 2000 and 2001. We are ramping up into this next maximum—it's an 11 year cycle. For folks who don't understand, when we talk about the solar 11 year cycle, it's simply [that] the sun is like the earth, in the sense that it has a north pole and a south pole—a magnet, a negative and a positive polarity.

But the sun does something a little weird: over 11 years it does a polarity reversal. Earth does not do that, obviously. And it is during the course of this polarity reversal when the magnetic field lines on the sun get so twisted and distorted and they push out, and the actual visible piece of that is a sunspot. When we see sunspots, like a meteorologist sees a low-pressure area, that's our nor'easter, when we see these big sunspot clusters.

So that's what the solar cycle is. And we've got periods, when you look at the left hand side, we are measuring the sunspots every month. And then things quiet down. The magnetic fields get settled and they ramp up again.

Now, the first key takeaway here is this: If you look at the next cycle, it's going to be a small cycle, perhaps the smallest cycle in the last 100 years. Unfortunately, some folks get the wrong message there and suggest that perhaps we are not going to see any significant activity for the next two years or five years or 10 years. That isn't the case at all.

And I'll just show you why that is. This is a plot of 250 years of solar cycles. I'm going to key in on two of the particular cycles: 1859 and 1921, and I know that will resonate with some people in this audience. The red, dashed horizontal line is average cycles. I'm pointing out those two cycles in 1859 and 1921 because both of those cycles were smaller than average and that's when the big space weather events occurred. When we talk about that Carrington event in 1859 and the potential consequence in the nation today, that occurred in 1859 with a smaller than average solar cycle.

So the takeaway here is, I don't care about the solar cycle. Yes, we are going to see less activity if I have fewer sunspots, but that big one could emerge, isolated rogue in nature. But it could sit there and produce that significant event. So do not dismiss space weather because it's a lower solar cycle.

It's interesting. People say, how do you guys know about 1859? We actually had the technology back then to measure the magnetic field of earth. So that very old green plot you see there is from an obscure observatory, of the magnetic field in 1859. So we actually have the records to see how strong the storm was.

The 1921 storm is interesting. I'm a bit of a history buff so I like looking back at *The New York Times* and some of the other publications, because needless to say, we talk about our reliance on the advanced technologies today. Well, in 1921 they talked about their reliance on advanced technologies on that day. And, indeed, even in 1921, space weather impacted the technologies of that day. If you look at the article on the left-hand side, from *The New York Times*, it will tell you that when the big geomagnetic storm in 1921 occurred, it said that the entire signal and switching system of the New York Central Railroad below 125th Street [was] put out of operation, followed by a fire in the control tower due to the induced currents.

So we've understood this phenomenon, this space weather, for quite some time. It's always had some impact on technology. But, of course, today, we're concerned it would have a tremendous impact on technology.

Now that's kind of difficult, again, to see in the back, but let me tell you what this is. It's always interesting. Our folks in Canada, Alaska, understand the Northern Lights—they see it quite regularly up there. Imagine what the folks in Cuba thought on that September morning in 1921, when they saw the aurora right before sunrise. That's how intense this magnetic storm was, that the beautiful aurora borealis was visible in Cuba, Hawaii, and other places.

Now, when I mention dodging the bullet, this is a really important message to take away. And we talk about this Carrington event in 1921 or 1859. I tell people, Carrington events occur—in every 11 year solar cycle, we see some Carrington events. We see them but the perfect storm from their perspective is that CME (coronal mass ejection), the eruption from the sun, has to impact earth. But the eruptions in the sun are occurring all the time.

This is a case in point. This is 23 July, 2012 and I believe Peter Pry mentioned it in the panel discussion this morning. We were sitting there and we were looking at the sun from the earth side and the yellow image here. There is a little sunspot there, about the size of the earth, but really of no significance. And all of a sudden—what you see on this other image, the blue image, this is what we call a coronagraph, where we essentially create an eclipse: we're blocking out the sun with an occulting disk. You see all that material shooting out? That is an extraordinary, fast coronal mass ejection.

That got our attention real quick in the operations center, when we saw something like that. Our first concern is: Where the heck is it coming from on the sun? Is it the front side of the sun coming towards earth? No, because the sun was very quiet—the sun looked like that. Fortunately, though, in the technology today—and our good friends in NASA provide a great service—where we have what's called a STEREO (Solar TErrestrial RElations Observatory) mission. STEREO by its name will give you some clue of what it is: two different spacecrafts in orbit around the other side of the sun so we can see all 360 degrees of the sun, which is great.

So now I said, "Okay, let's find out where this eruption came from." That big, bright area there was a tremendous flare. And we were able to measure material that is blasting out into space was moving at almost six and a half million miles an hour. So it could make the sun-to-earth 93 million mile transit in about 17 or 18 hours at that speed. The Carrington event in 1859 made that 93 million mile transit in 17.6 hours.

So by many, many measurements this event that occurred in July of 2012 was Carrington-like. I don't say absolute Carrington because we don't know how it is going to couple with the earth's magnetic field and how the induced currents will occur, but all the other stuff was there.

October 2003, again, on this theme of dodging the bullet, other things we've seen happen in the last 10 years—that people don't realize because haven't had too much of an impact—these were Halloween storms. They had impact, but it could have been much worse. We're looking at kind of a worst-case scenario and I'll tell you why. You see the big, big sunspot cluster there on the left-hand side of the sun. That's about 10 to 15 times the size of Earth. The sun takes a full 27 day rotation. When I see something like that, and my friends in the forecast center see something like that, we've got issue. This is kind of our nor'easter.

But what happened over the course of the next week here—and you'll see as this goes into motion—watch what rotates around the lower left-hand side: another monster sunspot group. And watch what develops above it, another—we had three nor'easters we were dealing with, all of them as big or bigger than what Carrington was looking at in 1859.

If this situation should occur today, we have processes in place now to notify Craig Fugate, FEMA leadership, National Operations Center in the DHS and the National Security Staff, and many others in government leadership. We recognize now what we didn't back then: the threat is so great. But this was a perfect scenario. We could have gotten nailed. We did get considerable impact, as I showed in that first slide, globally, how it impacted Earth: the power grid came down in Malmö, Sweden, there were damaged transformers in South Africa. We were spared. We were lucky. We had some big events, they just weren't any extreme events…but they could have been.

So, again, just a takeaway here that these are the big events that have occurred and we just got lucky. And I just wanted to finish—that big, big sunspot on the lower left hand side, as it rotated on the very right hand side of the sun, it was about to make its exit off—it had done its damage, it had done its piece—it had one last shot for us. On the fourth of November, and you'll see the image on the upper right hand side, there was another powerful blast. And that ugly trace on the lower middle there is the X-ray emissions, and that was the biggest X-ray emission we have ever seen since we started looking at these things about 40 years ago.

So that coronal mass ejection, shooting out to the right—it's not earth directed. Again, we got lucky. So just recognize, these things are happening quite regularly in the sun. But all the ingredients have to be there. And that last ingredient is that it has to be Earth-directed. Sometimes it's not.

Now I just want to touch on a couple of the space weather impacts. A lot of people say, "I'm just not getting this space weather stuff," so if I give you some idea of some of the impacts we see, you'll get some idea of why we monitor this so closely.

I was down in Atlanta last year visiting CNN. The CNN folks were asking, "Well, Bill, how are we doing with our coverage of space weather?" And I said, "You do pretty good, except you get something wrong every time." "What's that?" I said, "Every time when the storm is over, you say, 'Nothing happened.' You are wrong. Things are happening all the time." People don't want to talk about it, they don't want to share it, and a lot of it is really just not sexy news stuff.

We have a diverse customer base—imagine GPS, GPS pervades society today. We have so many users, so many critical applications of GPS. We get these moderate storms, and GPS users get impacted, I assure you. Even with a small solar cycle and not much happening, I just wanted to show you a couple of things, like in the March 2012 activity, the big LightSquared satellite zapped. They thought it would be out for maybe a day during the impact from that solar flare. It ended up being down for three weeks.

The DoD doesn't like to talk, needless to say, too much about impacts on our classified and other sensitive systems, but you'll see in *Stars and Stripes* that the general did come out and say, "Yeah, it interfered with the Air Force satellites. I'm not going to give you too much more info." But it will impact various systems.

Aviation always gets impacted. And if you read down here, 13 May, 2013, this is a report that comes from Air Traffic Control. We have nice relationships set up with these folks and they give us feedback. I know a lot of our folks in the audience here are either hams or like HF (high frequency) comm operations. Solar activity causes severe impact to high-frequency communications. These people are trying to get messages to aircraft to go up, to down, to go left, to go right and they cannot, because the space weather is impacting their ability to communicate. Again, we don't like to share this stuff. It is not stuff that is going to make the press, but it will happen every time we have any kind of space weather activity.

Here's another HF impact, just to show you. When we have a big solar flare—if you look at the green image and you see the big, bright flash, obviously, that's the solar flare. On the right-hand side, the image of the earth shows where the high frequency communications will be impacted. When the flare occurs, within eight minutes of that thing happening in the sun, we're feeling the effect on our ability to communicate right here on Earth. And when you look at this image, it's a sunlit side of the earth, and anything red is bad. Red means we've lost the capability completely. Lower frequencies would be degraded when we are in the yellows and blues.

Watch what happens. The flare goes off and all of a sudden, *whoosh*, across the Atlantic, the Pacific, HF comms are shut down within a minute. GPS gets impacted, radar systems. We get this information into the DoD within two minutes. They need to know what kind of interference is impacting their radar systems and other technology.

And I just wanted to show you an impact on the GPS. This was kind of a phenomenal impact and it will give you an example of what can happen during a solar flare. You see the big flare on the image on the left-hand side—very bright obviously. But on the map of Earth with the green dots, [the dots] are a GPS receiving network. Green is good. When it goes red, it's bad. We have this big flare—and this was actually during the solar minimum, which is why I point out space weather can occur any time, 6 December, 2006.

But I'm going to put this in motion because we get this big burst. And watch what happens on the sunlit side of the earth. When those green dots go red, we've lost that GPS capability. So you'll see the little red dot in the bottom, things are not too bad; we're not seeing any big bursts. And then all of a sudden, the spike here is the burst. The spike goes up and they are whacked, our GPS receivers—northern hemisphere, southern hemisphere.

Now, if you're driving down the road, relying on you TomTom or your Garmin, and it's not working for 10 minutes, no big deal. You'll get past it. But if you're landing a plane in a 200 foot ceiling in a quarter of a mile visibility with a crosswind in Juneau and you're relying on GPS and it's gone for a couple of minutes, you've got a big, big problem. So the FAA and many others recognize that. We recognize this threat of space weather to GPS, so we'll always have redundancy set in on some of those critical decisions like aviation.

This is a report that came in from the GPS Operations Center in Schriever Air Force Base in Colorado, just kind of verifying. I always like to show that these are the words not from Bill Murtagh and the Space Weather Center, but from the users, the people that rely on this information. And they said, "At 1930 Zulu UTC, widespread loss of GPS in the mountain states, specifically around the four corners." Aircraft were tracking six, seven, eight GPS satellites like they're supposed to. Within seconds they were down to zero or one.

Big solar radiation storms occur. And when they occur—if you are flying from the United States to Asia, what's the best route or the typical route these days? Polar. You go right over the North Pole. From point A to point B, it's the quickest way with the aircraft we have now that can fly that long. The biggest concern—there's no thunderstorms, tropical weather, obviously, or anything like that in the North Pole—what they are concerned about is space weather. It will impact communication systems, it will impact the navigation systems. And that little controversial piece: it also creates a biological threat, increased radiation exposure. The airlines don't want to deal with it.

As soon as they get the warning from the Space Weather Prediction Center, they say, "Okay, we're not going polar. We're going to avoid it." Last year there were 11,300 flights that flew polar, 13 major airlines. Those folks will start rerouting as soon as these space weather events occur.

So I'm just going to finish up just showing this. I mentioned it earlier. When we get a coronal mass ejection, a blast of material that shoots out into space, a magnificent explosion, but it's got magnetic cloud. It just shot out a magnet, and that magnet is going to interact with the earth's magnet and we are going to have a storm. We'll issue a watch, two days in advance sometimes, to give people a heads-up. It will impact lots of different technology. If John were here, this is where I would turn it over to him and say, "John, give us the power grid."

But I can give you the takeaway without the technology and the better explanation John would give. The bottom line is: when the CME impacts the earth's magnetic field, we are going to have a reaction. There are going to be electrical currents developed in our magnetosphere and ionosphere. They will manifest themselves right on Earth. We'll have current induced on the ground that is going to find its way often into big conductors like the power grid, pipelines, railway lines, and whatnot.

The bad news is this, folks: perhaps the most vulnerable place in the world is the United States—the northeastern part of the North American continent. We have all the ingredients. We have the very intricate power grid that acts like a big antenna. Obviously, we need that. We have the higher latitudes. The geology is critical—how the current will flow through different types of rock formations.

Our position relative to another big conductor, salt water, the ocean, all come into play. In the recent Lloyd's report, Chuck, remember, it said that the most vulnerable location in North America, was the corridor between Washington, D.C. and New York City. Not what you want to hear. That is the reality, though, and we recognize—even with the moderate storms we get—that indeed, it's those locations that do get impacted, the nuclear power plants in New Jersey and Pennsylvania and up in New England. They get impacted regularly. Manageable. We just hope it is during the big one. Thank you very much.

MODERATOR (CHUCK MANTO): Thank you, Bill. And we have one question. Identify yourself and ask a question for Bill.

AUDIENCE MEMBER: Meg Abraham and I've worked extensively on the satellites that track this. And I wanted to know—so we've got the GOES (Geostationary Satellite) system, for instance, that gives us a lot of early warning. How much warning are you able to then translate out to government entities? I know that the satellite doesn't get the information instantly, either. So there is a stepped or time delay.

BILL MURTAGH: Right. There's two pieces to that. The coronagraph that shows the material erupting from the sun—fortunately, that's the stuff that causes the geomagnetic storm. And there are several different types of space weather. That's the biggest, most critical one, and that's the one we get the most lead time on, because that is just stuff as it erupts from the sun. Unlike some of the other material, it's moving slowly. It is only moving about five million miles an hour. (It's all relative.) But it is actually making a 93 million mile transit. That is pretty slow.

So we can give the power grid first heads up, first forecast, first warning sometimes 15, 20 or even 50 hours in advance. But then we've got—and I didn't highlight this—we've got that L1 satellite. It sits in the Lagrange orbit, which is a million miles out into space, along towards the sun. So we've got this buoy out there, kind of like a buoy off of the shore of Miami. When the hurricane hits it, we get our first readings.

That's where we get our first readings. As that CME hits that spacecraft, now I know, now I'm dissecting it. I know what the density it, the temperature, the speed, and the magnetic field. So now I've got to issue an imminent warning because it's going to be here in about 20 minutes or 30 minutes. So that would be the process. So there are degrees of lead time. Of course, the accuracy would be another story we can talk about separately.

Planning for High-Impact Disasters
in Light of Recent Disasters

http://youtu.be/zmxhXymf4fg

Major General Robert Newman, former Adjutant General of Virginia
Dr. Paul Stockton, former Assistant Secretary of Defense for Homeland Defense

MODERATOR (ROBERT NEWMAN): I consider it a real honor to introduce our keynote speaker today. The Honorable Paul Stockton served as the Assistant Secretary of Defense for Homeland Defense for President Obama and just recently left that office. Dr. Stockton is a noted authority on homeland defense issues. He wrote extensively on this in his capacity as a senior research fellow at Stanford University. Dr. Stockton, other than the fact that we both cheer for the Cardinals, I have nothing to do with credibility and educational opportunities at Stanford. I wish I did but I don't. But I'm a big Cardinal fan and wish them the best as they conclude the season here.

But Paul has been a real advocate for the topics that we're discussing over the last couple days. And I can tell you, as an Adjutant General, there was no stronger ally we had in the Department of Defense than Paul Stockton. He has advocated for issues of concern to the states in his federal capacity, recognizing that it's the states that are the first responders, the ones that'll be in the mix first and probably for the longest period of time. And through his great efforts here, he has made our country stronger and has strengthened the ability of the National Guards throughout the 54 states and territories to respond to disasters, whether it's a local disaster such as a flood or a hurricane, something that's more typical, or certainly as we address these more catastrophic issues with things such as EMP (electromagnetic pulse).

So it's with great pleasure that I introduce my good friend, and a great American, Dr. Paul Stockton.

PAUL STOCKTON: Bob, thank you for those remarks. And thanks for everything that state National Guards across the nation do every day to provide for the public safety and security of their citizens. I want to thank Chuck Manto and InfraGard for hosting this event. And I want to thank all of you for focusing your energy, your attention, your intelligence on issues that don't always get the attention that they deserve, given the scale of the challenges that we have.

I appreciated the video clip there. It does provide some motivation. I actually take a different approach to these challenges, though. That is, in addition to being scared, I think it's very important for us to remember that if we collaborate together, there are very important opportunities for progress. So, rather than only imagine that the sky is falling, and that we're going back to the stone age, I think there are very actionable, very practical steps that we can begin to take, in order to build resilience against these kinds of hazards.

I would like to discuss two topics in particular. First of all, cyber issues, since they were the focus of the clip, but then I want to dive into electromagnetic hazards. Because I think that's a prime example of an issue that doesn't get the attention that it deserves. And I'm talking not only about severe solar storms, but of course EMP as well, because of the combination of E1 and E3 effects that are so challenging.

First of all, in the cyber realm, I think the administration and its industry partners are making terrific progress in building the kinds of voluntary standards for prevention and protection that the nation is going to need. And I applaud those of you who are helping advance that progress.

I see plenty of attention focused on the challenges of prevention and protection. I see very little focus on the challenges of response. And let me talk to you about the nature of this challenge, and ask for you to work together with me in making progress in this regard as well. Given advances in the threat, SQL (Structured Query Language, a computer programming language) injects everything that's going forward. The availability today of weaponized zero day exploits for sale on the web, in a completely unconstrained, wild west market, I believe that we cannot count on having perfect prevention, perfect protection, even with intelligent, well-informed investments in that realm.

I believe we need to make the assumption that someday, there will be a successful attack, either on the electric power grid, or perhaps on multiple infrastructure sectors simultaneously, that depend on common industrial control systems, control panels, everything else that's essential to the functionality of these infrastructure sectors.

Let me say it again. I think we need to continue to invest strategically in prevention and protection. But we also need to be mindful of the risk that someday, despite our best efforts, those measures will fail, and that we will be called upon, with the National Guard, with all of the partners represented here—federal, state, local, and industry playing a vital role—to limit the damage, limit the risks to public health and safety, limit the risks to national security and the American economy that such an attack could create.

There is a particular concern I have. And that is that the National Cyber Incident Response Plan, the incident management plan for cyber issues that provides the basis for going forward for response on an interim basis today, is built around completely different organizing principles. It would take the United States in a very different path than the National Response Framework that would govern the response to the physical damage that such an attack would create.

Let me tell you in particular what I'm concerned with. The National Cyber Incident Response Plan, the interim plan that would guide the response to cyberattacks, how to scrub malware, how to deal with the cyber-specific problems that a successful attack would create, is organized in a way that does not provide for a leading role for our nation's governors.

Our nation's governors are responsible, under our Constitution—they have the lead responsibility for the public health and safety of their citizens. To have a plan that doesn't put governors and state governments front and center for helping to manage the process of response to a cyberattack, I believe that's unrealistic, and I believe it fails to take advantage of core capabilities that need to be strengthened at the state level.

In contrast, we have a proven National Response Framework where governors play an absolutely vital role, where governors identify what kinds of capabilities to request from the federal government when state capabilities run short. That's an example of a vital role and very effective role that governors played in Superstorm Sandy, in Irene, in all the other natural catastrophes that we've faced in the past. I believe we need to leverage that terrific state-level leadership. I believe we need to find some way of integrating cyber-specific response with the kinds of management of the physical consequences of a cyberattack that does not exist today, and that, folks, we are going to need when—not if, I believe—there's a successful cyberattack on the infrastructure of the United States.

Let me emphasize again, though, that as we make progress in the response realm, it's critical that the voluntary standards that are currently being refined by the administration and its industry partners, and indeed, industry on a voluntary basis continuing to strengthen prevention and protection, that's also essential. Let's just not forget about this other policy realm that hasn't gotten the attention that it deserves.

Let me turn now to EMP threats, because it was a special area of concern for me in the Department of Defense. Of course, everything I say today is as a private citizen. I'm out of the business now, but I remain concerned that EMP hazards don't get the attention that they deserve, and especially that there are very actionable, very practical steps that can be taken to better understand the nature of the challenge, so that we can make limited investments in protection and also in response that can drastically reduce the potential consequences of such an attack.

Folks, the sky is not falling. It's not time to crawl into a cave; it's not time to play Chicken Little. It's time to think about intelligent, limited investments, eminently affordable, that could have huge benefits in terms of limiting the potential damage of such an attack.

Let me talk to you about a couple of particular issues that I think we need to further address. In order to provide for targeted investment of capabilities to prevent and protect critical infrastructure, I think, in some realms, we still need a better understanding of what kind of physical damage E1 and E3 are going to cause to the systems on which we depend. "Systems" include cars and emergency vehicles that are all laden with electronics and which are more complex than some of the vehicles on which past EMP tests have been conducted. How about the control panels for the emergency power generators that will absolutely be essential going forward, should there be an EMP attack, and components of the grid—high-voltage transformers and other substations, everything else that's required for grid functionality, based off of physical damage.

To what extent are the kinds of black start capabilities and other components that are going to be required to shorten the recovery period, but also provide for emergency power for those facilities that are absolutely vital to keep up and running, including chemical facilities, nuclear power plants, everything else that's required to limit the potential devastation of one of these attacks, I believe we need to do further analysis of what kinds of vulnerabilities exist, and how limited, targeted, strategic investment and protection can have the biggest bang for the buck. Resources are constrained. Let's be smart about how we invest, and I think that requires further research and analysis.

Secondly, I'm concerned that we haven't done enough to think about the environment within which we'd be trying to preserve the functionality of lifeline systems of infrastructure in such an event. Sure, we can all think about the immediate damage, for example, to the electric power grid that an EMP attack would create. But, if you think about what it's going to take to restore the functionality of the grid, that restoration effort is going to depend on other infrastructure sectors that are themselves going to be severely compromised by an EMP attack, either directly or indirectly, because of the loss of electric power.

Let me give you a couple of prime examples. Transportation assets are going to be absolutely vital if, for example, the industry's terrific current transformer reserve program, STEP (Spare Transformer Equipment Program), could be expanded in a targeted and intelligent way, so utilities can leverage their current, very strong abilities, to provide for additional transformer support if transformers in a limited area were destroyed by an EMP attack. These are big transformers, right? Some of them require special rail cars to transit. Others can be taken by trucks. But the ability of our transportation systems to function and enable those high voltage transformers to get where they need to be, that is going to be severely compromised by the same EMP event, and by the cascading failure of critical infrastructure that will result from the attack.

Same thing for communications. I am concerned, I'm concerned today, folks, that for a variety of threat vectors, we are not going to have the communications facilities—communications capabilities—that the nation is going to need in order to provide for the restoration of grid functionality, in order to get lifeline infrastructure back up and running. I'm worried that our communications sector is going to be severely compromised. I'm worried that utilities won't necessarily have access to the kind of reliable spectrum for network communications that are going to be essential so utility crews know where to go, know what kinds of operations are going to be conducted, and how to provide for the targeted restoration of power and other critical lifeline infrastructure in this severely disrupted environment.

This is an area for further analysis that I think is absolutely vital. That is, the cascading failure of the very infrastructure that we're going to need to get electric power and everything else that we need back up and running.

And then finally, I want to emphasize the need, going forward, to take actionable, concrete steps. My view is we've done a lot of admiring of the problem, and I've suggested some further ways in which we could admire the problem—for example, to better understand the physical vulnerabilities of critical infrastructure components to E1 and E3. But, even as we better understand vulnerabilities, I think it's important, as the State of Maine has done, as other states are beginning to consider to do, as the National Guard is doing across the nation, and of course, as key components of the federal government provide leadership, it's important to begin to think about what we can do in concrete, step-by-step ways to support industry, to help industry take the lead in protecting critical infrastructure against these kinds of threats.

So those are some very brief remarks. Because I'm a former professor, I'd love to speak ad nauseum, but I won't inflict that kind of torture on you. Instead, I'd like to hear your thoughts, answer any questions, welcome your perspectives, before we turn it over to a far more distinguished panel than I am. Yes, please. And if you could introduce yourself please.

AUDIENCE MEMBER: My name is Erik Gaull. I'm a lieutenant in the Metropolitan Police Department. I appreciate your comments, because it makes me happy to know that I'm not the only one who lays at night thinking about these things. Specifically, the thing I'd like to put out to the audience, to the people that think about these things from a far more scientific way than I, of the technical capabilities to solve these problems, law enforcement, the people, the emergency responders that are going to be called upon to try and restore order, keep things moving along, keep the scenarios that we saw on that video clip from happening, is becoming increasingly more dependent on pushing electronics and access to databases down to the field level, to the extent that we have mobile devices in cars.

The New York City Police Department has smart phone applications for its detectives to be able to query databases. You know, in the fire and rescue service, we're increasingly dependent on the ability to get information and transfer it in the field rapidly, so that we can get it back for what they call, you know, at the fusion centers, the common operating picture.

And so I think everybody is concerned about getting utilities back online. That's all well and good. But the pointy end of the spear needs to be protected as well. And more thought needs to be given to the dependency of the first responders on the system.

PAUL STOCKTON: I completely agree with that. We're going to need those first responders. We're going to need their contributions. We're going to need the reliability of their communications, all of their operational capabilities, in this kind of scenario, in a way that would be unprecedented.

MODERATOR (CHUCK MANTO): Do we have another question?

AUDIENCE MEMBER: Dr. Stockton, we met two years ago in Aspen. And you spoke, at that time, about plans in the Department of Defense to build power plants on bases and wield power to the private sector. I've lost track of that. Where does that stand?

PAUL STOCKTON: The opportunity for the Department of Defense to be able to ensure that it can still execute its critical missions by virtue of having more reliable electric power, in part, potentially, by having power generation on military facilities, and providing for public-private partnerships to provide for critical loads on the military bases, that opportunity still exists. It's going forward, but maybe it's not going forward as fast as the threat would require us to make progress.

MODERATOR (CHUCK MANTO): One last question.

AUDIENCE MEMBER: My name is Larry Guinta with AZ-Tech in Scottsdale, Arizona. Mine is kind of a statement back to the first responders. Our group has worked on some patented technology, we've reached out to first responders multiple cities. We've responded back to multiple cities to see if we could assist and let them be aware of briefings on some of these technologies. And too oftentimes, somebody at the desk or a secretary says, "We have a consultant handling that." Or, "It's already being addressed, and we're not interested." And it goes no farther than that. So, if anybody here is a first responder type, you need a key buzzword, that if somebody calls offering solutions, I believe, to make sure that rises the top, so somebody can make a realistic decision about if it's really valid or not.

MODERATOR (CHUCK MANTO): Thank you very much.

Panel—Planning for EMP and High-Impact Disasters

http://youtu.be/6fFSh7cJieU

Dr. Richard Andres, Chairman, Energy Security Program, NDU
BG (NYG-Retired) Kenneth Chrosniak
Ambassador Henry F. Cooper, Chairman, High Frontier
MG (Retired) Robert Newman, U.S. Army retired and former Adjutant General of Virginia
Dr. Paul Stockton, former Assistant Secretary of Defense for Homeland Defense
CAPT James Terbush, MD, USN and U.S. NORAD/NORTHCOM, J9

MODERATOR (RICHARD ANDRES): This is a really, really tough problem to solve. It's owned by so many different folks out there that getting ownership for any particular organization is difficult and solving the problem is absolutely tough. What we're going to need to do to solve this is begin at the bottom level with individual citizens. Movies like the one you saw I think are very useful for reaching out to the average person at home. It gives them an idea of the importance of something like this. We need to look at the municipal level, the county level, the state level, and the federal level. But we cannot count on the federal level solving this problem. This is too big, this is nationwide. We're going to have to have a multi-pronged approach to finding solutions to this very, very important national security problem.

And in terms of doing that our panel here is well situated to come up with some ideas. And just briefly I'm going to introduce them. General Ken Chrosniak, Sir, thank you, Ambassador Henry Cooper who has been very active in this in many ways for many years, General Robert Newman, sir, if you could, and of course you have heard Paul Stockton and Captain James Terbush, Sir, thank you very much. And with that I'm going to turn it over, and I think we're going to start right over here with General Chrosniak.

KEN CHROSNIAK: Thank you, Sir. It's a real honor to be up here with the panel that is here, and especially with my idol, Dr. Stockton over here. I remember he came to the Army War College to the DDE (Department of Distance Education) class of 2012 and gave on the 23rd of July, I remember—I have your speech memorized—when you said that EMP (electromagnetic pulse) keeps you awake at night. Secretary Panetta had asked you to come over there and talk about the Title 10, Title 32, and the New Madrid Fault and so on. And that really perked our interest quite a bit at the Army War College, because it raised it to a different level with our Commandant.

The bottom line up front, I was asked to talk about EMP (electromagnetic pulse) preparedness for the military, our survival. And I agree with Dr. Stockton that response is the big, that's the long pole in the tent, to use a military term, and for emergency or regional catastrophes, and they are catastrophes—if you owned a bar in New Orleans, it was a catastrophe when it was under water—but as a regional catastrophe, the U.S. military is admirably prepared to do that. We have well-trained, -staffed, -exercised plans with NORTHCOM (U.S. Northern Command), ARNORTH (U.S. Army North), National Guard Bureau, they do a very excellent job and we have seen that. However, the military is not prepared for a catastrophic, cataclysmic event such as if it reaches that level we just saw a few moments ago.

An *ALBISA?* [03:03] is a maximum complex catastrophe, whatever you might call it, but I call it a cataclysmic event, almost a tragedy. But as Dr. Stockton said, we can attack this elephant and bite it off, bite by bite, and actually walk it down to the base and mitigate as much as possible. You're not going to mitigate everything, but you can, like Congressman Frank said, you can take it in small bites now and hopefully have some impact to save—we're going to lose a lot of lives anyway—but save some lives, quite a few.

I'll focus on three areas quickly. With my few years in military service I have noticed that the most important things that I have picked up in my few years of studying this, not as much as the other gentlemen here have been studying it, is basically on troops, training, and communications. First one is troops: you need people. You need people identified. You need them trained and so on. And we do not have military personnel right now probably within NORTHCOM, within ARNORTH and so on, we have wonderful CBRNe forces, you know chemical, biological, radiological, nuclear enhanced forces out there that are very good at what they do. The State Adjutant Generals, General Newman had the Human Home Response Forces and so on, and they're excellently trained if they can get out the front gate, but there aren't that many of them.

What you're eventually going to need when, as Dr. Stockton said one time, when the state governors call FEMA and say, "Send us all you've got." And then after a while they've sent them everything they have got, and then the FEMA (Federal Emergency Management Agency) Director, Craig Fugate contacts the President, the President calls the SecDef (Secretary of Defense) and says, "Send me everything you've got." Eventually we're going to need general-purpose forces. General-purpose forces will not be trained to go into this no-notice pick-up game. Some of you have been on some football and baseball teams when you were younger, just had a little pick-up game.

But you have to have what I call training now. You have to be able to at least acclimate and make these forces aware of what kind of environment they're going to be going into. Now the CBRNe forces are trained for this and some of the Home Response Forces are trained, but you're going to have untrained military forces going into multiple regions, not just one region, maybe all 10 regions, maybe four or five regions, but what do you do in the military, you always plan worst case situation, so all 10 FEMA regions are involved. Let's just say it's a cascading effect like the National Geographic *Blackout*.

All regions, all at the same time, over vast urban and suburban areas, multiple jurisdictions for a very long, prolonged period of time, maneuvering within a humanitarian disaster environment with overwhelming mass casualty, graves registration, sanitation, medical needs—without the benefit of the defense industrial base, because you're going to need those critical, just-in-time logistics providers and so on to be able to bring products and so on into the defense industrial base to be able to resupply the military. Coms resulting very little situational awareness, which is essential for the GPS for those forces—your position, navigation, and timing. A lot of the radios are synced to the GPS synchronization and so on. Loss of air traffic control is another issue.

I can go on and on, but you see this all within a radiologically and HazMat-contaminated environment. Being a first responder myself, the fire company and rescue and Vice President of the ambulance company in Carlyle, small town, we still run 8,500 calls here, even in a small town, but you're going to be overwhelmed. And as a military commander, you have to be able to sync that in with FEMA.

And FEMA, Craig Fugate about a year ago put out a wonderful scenario called "Critical Communications During and After a Superstorm" and basically it boils down to this: after 30 days everything, even the plain old telephone system, is down, except those that are HF, Mars, ham radio operators and so on. As long as you have power for your generator you will have HF line of sight, but we in the military, we deal BLOS—everything is beyond line of sight. It goes through satellites and so on.

The other one was training. Start training people—at least develop plans and scenarios for a nation-wide magnitude, of which we have no plans right now, and start exercising them, and get a lot of the smart minds in here to incorporate some of their ideas on how we can at least respond and hopefully mitigate the loss of lives that are naturally going to occur anyway. Sir.

MODERATOR (RICHARD ANDRES): General, thank you very much. Ambassador Cooper.

HENRY COOPER: I want to link back to the first panel this morning just a little bit with what I say. I want to address two issues. One is the threat from short-range missiles off our coast, including from the South, as well as the satellite launch that was mentioned by Jim Woolsey. The good news, as far as I'm concerned, is the Defense Authorization Act, which is under consideration now and is supposed to be passed in the next couple of weeks, if the Senate gets around to passing it. It will instruct the Secretary of Defense to come back with a program recommendation for how to improve the East coast defense, including against threats from the South.

And it's my belief that the Department of Defense below the Secretary is prepared to deal with this question. So, I believe the DoD may address many of the issues of concern over the next year or so. It's appalling to me that they have let it go this far without dealing with short-range threats, since we've known about it for well over a decade. And there are options.

There will be a lot of talk you will hear about needing a new site for missile defense along the East coast, extending all the way to Ohio, which is one of the places being considered. I'll let you think about why that is. I just want to make an observation that the Aegis ships that we have deployed—and last year any day at random would have found four to six operating or in port along our East coast—have the inherent capability of shooting down ICBMs (intercontinental ballistic missiles) that come at us over the North Pole. We demonstrated that by shooting down a satellite in 2008, I believe it was.

So, all that is needed is a radar up in Maine to make this happen. If you move the one that exists, and I forget what Army base it is, to Maine, it would cost like $20 million. If you bought a new one it would cost like $300 million. That's a lot of money, but chump change in the broader scheme of things—and certainly far less than building a new site. And we could, overnight, have defenses for those of us who live along the East coast from threats from the North.

Our ships can also defend against short-range missiles if they are nearby, and that is the case when they're in ports as well. So it's a matter of training in that case, and the Navy can certainly provide that capability if, in fact, they are called upon to do it.

The problem, a key problem is that none of our ships operate in the Gulf of Mexico, so we are wide open to threats from the South, whether they come from vessels off the coast there or from Venezuela, which has ties, as you may know, to Iran and other places. And the other thing that was mentioned was a satellite threat. If the Iranians and North Koreans choose to do, it they can overfly the range capability of the defense interceptors that we have, and we have none directed toward the South now and the ships can provide interim capability for that.

We need, ultimately, I believe space defenses to deal with that, but that is years away for other reasons. Enough said about those.

The second issue I want to come back to is the theme that was talked about by Ken at the end. I believe training is a key issue on the response side to deal with this problem—training beginning with the local responder community. And I believe the cohesion that the National Guard can play, including linked back to the missile defense world, where they man the sites in Alaska and some of the command centers around the country. So, the basic infrastructure exists within the National Guard that is linked into NORTHCOM and the federal level, and the Guard itself is made up, as you know, of firemen and police and doctors and whomever when they're not in Middle East and wherever and they're working from home, so you have the inherent ability of integrating the local responders within the Guard if we simply prepare. And I believe training is needed.

In South Carolina I am working with the Deputy Adjutant General there, and have been working for a number of years, and we're intent to put together an exercise. If this meeting had been week after next I could have told you a little bit more about the plan, but it's our intention to involve the inherent capabilities in South Carolina and neighboring states, perhaps extending to all of FEMA region four which is headquartered, if you're familiar, in Atlanta. And with Bob Newman's help (who knows the Adjutant General in South Carolina), maybe we'll come to Virginia and even Baltimore as we think about a scenario to be included in one of the normal National Guard scenarios, Vigilant Guard 2015.

It's my intention, at least, to encourage thinking between now and the execution of that exercise with intermediate steps that can be taken to help train the community of people right now who have—maybe it's not zero understanding, but it's closer to zero than what is needed—understanding of what they will be called upon to do if we have a blackout situation from any number of threats. Thank you.

MODERATOR (RICHARD ANDRES): Thank you very much. General Newman.

ROBERT NEWMAN: Thank you, Rich. It was said by some famous American that all politics are local. Well, you've heard it also said that all disasters are local, and those of us that have experienced the need to bring disaster relief to our fellow citizens understand the meaning of this. When you talk about first responders, or what I call "first responders of consequence," those people that can bring a lot of stuff to help relieve an emergency situation, you're talking about the National Guard.

Now, we're often guilty of assuming that everyone understands that the National Guard is a dual-headed organization, and you hear terms like Title 10 and Title 32 thrown around. Let me simplify it for everyone so we're on a common sheet to begin the discussion. The National Guard has two commanders. During times of peace and when they are not mobilized under federal orders, their Commander in Chief is the governor of each state, or the Commonwealth of Puerto Rico, or the territories. We have 54 TAGs, 54 National Guard units. In that capacity in the Title 32 role, which is U.S. code Title 32, the Guard is funded for all its training, for all the equipment that it uses by the federal government, but it takes all that training that is used for the war fight to go to Afghanistan, Iraq, or whatever a Title 10 mission the President will direct the Guard, it uses those capabilities, those trained resources, those Guardsmen, airmen, and soldiers to do whatever the governor determines that they will do.

In the other role, and this is where most of us think we know the National Guard, you'll see them in Afghanistan, you'll hear about a Guard unit returning or one that is departing or one that is serving over there. We think that the Guard is the reserve of our country and they're over there fighting the fight against the bad guys. And that's true, but in this context and in the discussions we have had over the last two days, I think it's obvious that the National Guard can play an important role not only in preparing to defend against a catastrophic or, as Ken said, a cataclysmic attack, but also to help recover from one, which I think is the real challenge, as Secretary Stockton was talking about.

I can tell you now that at least from my time, and I've been out of uniform for three years now, the National Guard was not ready for this mission. We had not planned; we had not trained; we had not exercised to the degree necessary to handle a cataclysmic event. Now, sure, we had units that could do that, and we had partnerships with other states through emergency assistance compacts, we had partnerships with the U.S. Northern Command and others. But on an individual basis I can tell you that the Guard was not there.

And we have seen how difficult it is on this mission of preparation for a cataclysmic event, whether it's an EMP or a cyberattack—something that puts us out of normal society for more than 30 days, let's say. The feds just don't seem to get it. Or, as Dr. Stockton said, or maybe it was Rich or someone, this is hard to do. Rich, I think you were talking about that. There are a lot of cooks brewing this stew and a lot of people with money and power, and of course that not only breeds conflict but an honest discussion with seldom results in good success.

I would like to suggest that the National Guard is one of the vehicles that this group should embrace to help us prepare for a cataclysmic event and then to respond for one. Here are a few reasons why I think we should do that. First of all, 90 percent of the time, the National Guard is under the command of the governors. The governor is a powerful political elected leader. He has immediate call in to the President. He has immediate conversations and regular discussions with our co-del, congressional delegations if you will, the Senators and Representatives from the state. He has through the National Guard relationships with the Department of Defense. He has through his Homeland Security Advisor relationships with DHS. And it goes along from there.

And on the state level he has access to a number of important agencies, such as his fire programs people, his Department of Emergency Management, those people that actually plan and operate the emergency forces during a disaster. He can access these folks, sell his ideas, the necessity of a properly prepared defense and then response capability if he chooses to do so. I think that the mission of this group should, through the Guard and through the governors, just as Secretary Stockton suggested, build from the bottom up the message that we need to get out and prepare better and to prepare our forces to respond. And I think that if we take that away from our meetings here over the last two days we will have accomplished a great deal.

I will conclude my remarks by saying that Ambassador Cooper and I are going to visit with the Adjutants General at their February meeting here in Washington a couple months from now and we plan to discuss this to start the ball rolling. And I think that from the response that I have received from the few TAGs with whom I have spoken, there is a great recognition that this new mission needs to be embraced more boldly than it has in the past, and I predict we will have great success.

MODERATOR (RICHARD ANDRES): Very well. Captain Terbush, Sir.

JAMES TERBUSH: First of all, I wanted to say thank you to Chuck Manto and InfraGard for inviting me today. Thanks also for including me on such a distinguished panel. I get to work at NORAD and U.S. NORTHCOM. I work in the Science and Technology Directorate, and this is one of the projects I have had the opportunity to spend a little time on.

I want to make three quick points today. Let's see, I have five minutes and 10 slides, so it should be a piece of cake. Okay. Number one, the health consequences of these threats—EMPs, solar storms, and also cyberattack—are significant, and significant for health because of the specific vulnerabilities that health critical infrastructure, the public health and medical sector, are subject to. Second, the health sector is not fully prepared, and we could go on with numerous examples, but I think Superstorm Sandy will highlight particular healthcare vulnerabilities and some lack of preparation. And the third thing, the third very quick point that I hope to make is that both of these vulnerabilities require change to policy—non-material and material solutions, so changes to infrastructure, hardware, if you will, alone are not going to do it.

So, how did we get started? Well, in conjunction with the Johns Hopkins School of Public Health and also the University of Pittsburgh Center for Health Security, we wrote a couple of papers on this subject. And I will go ahead and make reference to this book by William Forstchen, *One Second After*. I describe it as a terrible book, but in the context of a terrifying terrible book. And so it had, and I am a public health doctor by training, so it had three public health scenarios through my reading.

The first one of course was the effects on people who are in hospital or in long-term care facilities or rely specifically on life-saving medication. That group is the first to succumb to such an event or such an environment. The second group had to do with considerable loss of life associated with social unrest. And the third had to do with the society coming back to the caring capacity. But the first threat, the first public health die-off, if you will, was of particular concern to me.

So, I mentioned this collaboration with Johns Hopkins and UPMC (University of Pittsburgh Medical Center), and we did a literature review, and it turned out that there really has not been much written about this subject. The millennium bug was good for a few articles, and then in 2003 there was also another article about health consequences of a cyber event, but really this is an unplowed field.

Many of you have seen this slide before. It talks about the vulnerabilities and interdependence of critical infrastructure. I would highlight that both civilian and military capabilities rely upon these infrastructures, and Dr. Stockton has made this point many times that for over 90 percent of the services to military bases, and so that would be energy, sewage, food, transportation, etc. we rely upon the civilian sector, just like you do. And I would particularly also include the fact that our telecommunications, our IT backbone, we ride on civilian infrastructure just like the rest of you.

A series of cascading failures could interrupt some mission-critical functions of the military, and this was in essence the beginning of this project. We tried to link some of the mission-critical functions that our boss, General Jacoby has here in the homeland, and that would include such things as force protection and by extension force health protection. Well that gets into the issue that if General Jacoby doesn't have healthy troops to send in to respond to a disaster, there will be some lag. So, how can we improve the resilience of military medical facilities? That is the essence of this project. So, health is a mission-critical function, and it turns out that military healthcare facilities are just as vulnerable as civilian ones.

This one shows there are 16 now, 16 critical infrastructure sectors. It was mentioned earlier that they rely upon each other, but that healthcare and public health is particularly reliant on power and IT. It turns out that the way we practice medicine now in the 21st century has come a long way from even 20 years ago, and now particularly information technology is embedded in pretty much everything we do in health and medical.

So, these are some of the health consequences. And I'll run through this list very quickly, but there is a requirement now through the Affordable Care Act that we do electronic medical records basically for all patient encounters. The medical supply chain and the just-in-time delivery system is absolutely reliant upon web-based systems, pharmaceuticals likewise, billing and administration. We had a friend of mine go into the hospital not too long ago and was asking for immunizations, and she was told, "I'm sorry, we can't give you or your kids their shots. Our computers are down." So, we are very reliant on this. Imaging, lab results, certain medical equipment, health communications, emergency responders, this was mentioned already, are increasingly technologically driven. And finally, as you can imagine, a dark, hot hospital is a very unsafe place to be.

These health consequences may be mitigated. Most hospitals have a power system. Most hospitals will have auxiliary power in the form of generators. But as we saw from Superstorm Sandy they would tend not to last more than about three to five days, if for no other reason than the requirement to refuel. But these systems also need to be maintained. And I'm glad to see that the EMP SIG looks at significant threats besides solar storms, cyberattack, and aging infrastructure that could bring down the grid, such as pandemic.

I have one last slide here. Let me just go ahead and say that military is not immune. We have some hardening, if you will, but definitely need collaboration, cooperation, communication with the private sector, and transitioning some of these mitigating projects that we have started in the military to help make bases more energy resilient is the way to go. So, I'll leave it with public/private partnerships as the key message. Thank you.

MODERATOR (RICHARD ANDRES): Thank you very much. It's very difficult to have this many distinguished speakers talk about an issue of this level of importance in the time allotted, but the good thing is that now you have an opportunity to ask some questions. So, I would like to open this up. If you have questions please raise your hand, and say who the question is addressed to. Yes, Sir.

AUDIENCE MEMBER: I'm Alan Roth and it's addressed to whoever dares to answer this question. We talk so much of a disaster occurring based on let's say cyberattack, EMP, and other things, but I haven't heard much about multiple faceted attacks. If China has infiltrated our infrastructure through the internet the way I've heard they have and they are just waiting for the right time, wouldn't that right time be when we're having a heat wave or a deep freeze or a major nor'easter coming up the coast or other really bad things going on at the same time? How much planning, how much thought goes into your work that includes this factor of having multiple things happening at the same time?

PAUL STOCKTON: I think that's a very important consideration, and it's been taken into account for some time. I believe it needs to be considered even further. So, you have mentioned the possibility that an adversary would have ample warning of a storm approaching the United States, such as Sandy, but I would say we also need to think about a combined arms attack, that is the possibility that a cyberattack would be supported by a coordinated kinetic attack with improvised explosive devices or some other kinetic means on critical components of infrastructure. To have those two events occur simultaneously could present exceptional challenges rather than one threat vector alone.

HENRY COOPER: I would like to add the thought, and to remind you, that during Cold War we worried about Soviet Union attacking the United States, we planned literally to deal with an EMP attack as a precursor attack on the United States, basically with an attempt to shut down our ability to communicate with our strategic forces. And when I say "our" I mean supporting the President's decision. We spent an awful lot of money on hardening that system against electromagnetic pulse. We didn't spend a penny on hardening the critical infrastructures to support the homeland defense mission for a whole lot of other scenarios.

I hope that the current Department of Defense continues to support the testing and the surveys and all the rest that are assured to maintain, needed to maintain a hardened system of communication and our strategic forces. But I have to tell you that we have shut down a lot of the testing vehicles that we used during Cold War to save money, or for whatever the other reasons are. So, I worry about whether or not we have maintained even the capability that we worked enormously hard during the Cold War to achieve to deal with a precursor attack.

And as was stated here, if China or someone else is going to use this EMP, if you will, as the primary mechanism I think you can surely count on cyber and other activities as a prelude.

MODERATOR (RICHARD ANDRES): We have a question in the back here.

AUDIENCE MEMBER: Frank Turner. I was wondering if the panel would comment please on we have heard the perspective from DoD, we have heard from the State, can you talk a little bit about what you think your specific roles and responsibilities would be, especially all the way from prevention through response, and how you would pass command and control back and forth?

MODERATOR (RICHARD ANDRES): Sounds like a Guard question, but I'm not sure.

ROB NEWMAN: Frank, I think the Guard, as I had mentioned earlier, has a responsibility to plan for all contingencies, and this is a unique disaster that awaits us, so we need to do better planning, better interaction, the whole of the government approach used in the OCONUS (outside the continental United States) discussions now has applications to the states, too, in this. We need to train with our fellow responders, whether it's at the federal level or at the state level, and then we need to exercise to make sure that we can press to test and the circuits are connected.

I think the key here is this is not business as usual. The events we are discussing here over the last couple days are events that are going to task perhaps not a state but the entire region, maybe even the entire country. If we're not prepared for this in significant training exercises, we are sure not going to be prepared to take it on when it actually happens. I think we need to have a different paradigm when we discuss these types of emergencies and develop training exercises to ensure that we have the capabilities to do that.

KEN CHROSNIAK: And that's very true because the military does really great at planning for the last war, and you stay within your own comfort level, like a good Staff Officer will, but this is really thinking out of the box right now, so this will open up a whole new paradigm shift of planning. So this is going to be unique, but it's got to come from the top down. Or, like I always say as a Combat Engineer, the best way to take a bridge, this is out of *A Bridge Too Far*, I think, the best way to take a bridge, according to Robert Redford, was to take both ends at the same time. You attack it from the senior leadership, the 30,000-foot level, and you attack it from the tactical level of people, uprising within the people to contact their first responders and say, "What are you doing, National Guard Unit Commander? What are you doing to protect us from an EMP? What are you doing?" Then have it all the way go up through the FEMA, Fusion Center, and work its way up and work its way down.

MODERATOR (RICHARD ANDRES): We have a question in the middle of the room.

AUDIENCE MEMBER: Donald Donahue from the Diogenec Group and the American Board of Disaster Medicine. Dr. Terbush touched upon healthcare, medicine, and in healthcare there are hyper logarithmic advances in technology and as federal policies that are basically pushing us, it goes beyond critical infrastructure, it could be down to the individual. My pacemaker, I don't have a pacemaker, but my pacemaker could be remotely controlled. The broader panel, how do we address that? How do we address what becomes a very individual and broad-based technological challenge? I can't protect my hospital. How do I protect the 10,000 people that I support?

JAMES TERBUSH: Thank you for that question. There are numerous examples of medical equipment that are vulnerable to hacking, if you will, cyberattack—pacemaker you mentioned, ventilators. The one example I would use is that Kaiser Permanente, that has a footprint all over the country, bought some 8,000 IV pumps, and it turns out that these new sophisticated IV pumps can be programmed remotely from someone at a terminal. Their IT security person at Kaiser Permanente, to their credit, decided, "No, we're going to go ahead and instead fat finger every one of those just to obviate that possibility."

This comes under the area of cybercrime and those are issues, but they tend to be one-at-a-time issues, or could be one-at-a-time issues. The more significant thing is that medical equipment, pretty much across the board, has a computer chip in it, and if there were something, such as an EMP or extreme solar weather, it could affect a majority of those devices.

MODERATOR (RICHARD ANDRES): We have time for one more question before maybe we give Dr. Stockton or anyone else a last word, because we have literally like one minute. Identify yourself.

AUDIENCE MEMBER: Tom Popik, the Foundation for Resilient Societies. This is for Captain Terbush. You mentioned something that is very intriguing, which is the carrying capacity of the continental United States. There is some disagreement, I think, that if we were to have the electric grid go down and not come up what would be the casualties, and there have been estimates all over the place—two-thirds, 90 percent. It sounds like you've done some kind of a more detailed study, and I wonder if you could tell us about that.

JAMES TERBUSH: Well, a friend of mine taught me the use of the word dystopia as opposed to utopia…these dystopian scenarios, no, I haven't studied a great deal, but it turns out that the population of the United States, say in 1850, before we had benefit of a power grid and certainly before computer chips, was significantly less, probably less than a third of our current population. I don't know where you take that number, but in the 1850s we did know how to do things that we don't know how to do now.

MODERATOR (RICHARD ANDRES): Shall we open it up for any last closing comments that you would like to make?

PAUL STOCKTON: Chuck, let me thank you, InfraGard, but everybody in the audience, this constitutes the coalition of the will and we are willing to make progress. Let's stick together and continue to not treat this challenge as something that is so large that progress is impossible but that chunk by chunk, we're going to do what it takes to help defend the nation.

HENRY COOPER: Yeah, I have one thing I left out, because I was closing on my allotted time. I mentioned that we were vulnerable to threats from ships off the coast in the Gulf of Mexico, and couldn't right now do anything about it. We can deploy the same kind of defenses that we are putting in Romania to be operational by 2015 and in Poland by 2018 in military bases around the Gulf of Mexico, and I'm working with the local folk and universities and the state governments to try to make that happen.

KEN CHROSNIAK: I do just have one more thing. I just have to say this. I think the soldiers, the troops, I'm thinking the responders, it's actually the carbon-based unit that has got to show up and actually go out there and do the recovery and so on and help his or her fellow citizens, and that's where the long pole in the tent is. I talked to a Battalion Commander who got back from Iraq not long after I did, and we were discussing EMP. He is a local Battalion Commander in the state of Pennsylvania—I certainly won't mention his name—who said, "I am not even going to go to the armory. I'll be home with my family where I belong." And that came from a highly decorated Striker Battalion Commander. And it really struck me then—this is about a year ago, right around the time of the NDU (National Defense University) exercise. That's the way they're going to think unless you give them some hope, give them some training—they think they're going into an overwhelming environment. I don't care if it's National Guard, Army Reserve, General Purpose Forces, or CBRNe, you are going to go into an overwhelming environment, an unknown environment.

If you start training, make up scenarios which should have been done a long time ago, from the DHS and so on—should have been a long time ago—the 2008 EMP Commission report has been out for five years now. What has been done? That's what the troops need, the response forces.

Panel—Cost-Effective Power Grid Mitigation in 2014

http://youtu.be/zDlZcruZ30c

Mr. George Anderson, Founder and Chairman of the Board, Emprimus
Dr. William Joyce, Chairman, President, Advanced Fusion Systems
Mr. Gale Nordling, President and CEO, Emprimus

MODERATOR (CHUCK MANTO): The folks up here have been so concerned about this that they all in their own unique ways have put their lives on the line and have sacrificed substantially to make technical and practical solutions that are compellingly affordable.

Dr. William H. Joyce is the chairman and CEO of Advanced Fusion Systems (AFS), the retired former chairman of the board and Chief Executive Officer of Nalco Company, leading provider of integrated water treatment and process improvement services, chemicals and equipment programs for industrial and institutional applications. He served in those roles from 2003 to 2007.

From 2001 to 2003, he was chairman and chief executive officer of Hercules, Inc., a global manufacturer of chemical specialties. Prior to that, chairman, president, CEO of Union Carbide from 1996 to 2001. In 1999, presided over the sale of Union Carbide to Dow Chemical, and served as vice chair of Dow Chemical until his departure.

So, right now, we're very honored to have Dr. William Joyce begin his presentation, which will give us an overview of what we can do practically about these problems. And then afterwards, I'll introduce George Anderson and we'll go from there.

WILLIAM JOYCE: Thanks, Chuck. I appreciate all of the work that you and others have done to put this on, and thank all of you for being here.

AFS is a new company, using new technology, previously tested in military applications, and we offer protective devices and sources for EMP (electromagnetic pulse) and GIC (geomagnetically induced current), and GIC and EMP test services. Here is a problem that we all have: getting people to listen to advice about terrible things that may not happen for years.

The tsunami stones are 127 feet above sea level. So, now comes time to build a nuclear plant just below these stones, and the first thing they do is dig out the ground to get closer to sea level. So, I think that's a problem that faces all of us. But it's a challenge we're going to have to find a way to overcome.

I want to talk a little bit about EMP and non-nuclear EMP. The best available technology that we have now is MOVs (metal-oxide varistors) and lightning gaps. Depending on these devices is questionable in a nuclear EMP environment, and unacceptable in a non-nuclear EMP environment. And the following graph sets some current technologies and AFS technology against the threats in a scaled view.

The dark red line there is an EMP pulse from a nuclear explosion. The light red is a non-nuclear, and of course, lightning is shown there. This is a log-log scale, so what seems like a close comparison between two is really an order of magnitude more dangerous. So, the top of that graph has a lot of threats to it.

The blue line shows where MIL-STD-188 devices would give protection. The green line there is where an AFS Bi-tron gives protection.

Let's talk a little bit more about non-nuclear EMP. Numerous powerful non-nuclear sources have been demonstrated, some have been certified by the U.S. government having field strength in excess of 250 kV per meter. Some of these devices, including the big ones, are portable. This presents a threat that exceeds the protection that you would get from MIL-188-125. These threats are relatively inexpensive to make and lend themselves to multiple simultaneous attacks at different locations.

This is a 35 kV per meter system that was built in collaboration with the U.S. Army, and I might add the more modern and powerful ones fit in the same kind of van. So, they're really relatively easy to build one of those and drive by your favorite substation and take it back to the eighteenth century.

MIL-188 is the current government standard for high altitude EMP protection. It was good at the time, but it needs updating. The utilization of a swept narrow band source for shield effectiveness ignores the physics of the response of materials to an ultra-wide EMP pulse. The slow response of MIL-188 designs allows too much energy leakage before the protective device becomes capable.

AFS offers some testing services that exceed MIL-188 and gives us a realistic test environment for nuclear and non-nuclear environments.

I just mentioned our devices depend on a new version of electron tubes. A few words on how they differ from semi-conductive devices: we use tubes that were originally developed for military EMP simulation and high-power microwave applications. They're designed for repeated operation in this extreme transient environment. So they're very robust. Solid-state devices can fail by piezoelectric-induced overstress, leading to single arc failure, thermally induced overload and cascade failure. The AFS tubes are significantly faster than the fastest power semi-conductive devices. And the tubes are not subject to DV/DT (change in voltage over time) or DI/DT (change in current over time) constraints, as are semiconductors. Typical slew rates are well in excess of megavolts per microsecond.

For DC operation, we use a Pulsatron. It's a high-vacuum, cold-cathode triode tube. I won't go through all of the specifications there, but you see despite its small size of 5 inch diameter and 12 inches long, it can handle 500 kV and 250,000 amps.

The Bi-tron is what we need for AC systems. It's a bi-directional electron tube designed for AC power, electronic switching and control operations. Voltage ranges to 1.2 million volts and hundreds of thousands of amps. For over-voltage protection, including EMP protection in excess of MIL-188, the Bi-tron is designed to operate in EMP environment, and is capable of handling repeated pulses at a multi-kilohertz rep rate. Size varies with voltage, as you would expect: units below 35 kV are less than 16 inches in diameter and no more than 36 inches long. Units for 1.2 million volts are approximately six feet in diameter and their length is basically the same as you would see for an insulating support for high voltage applications.

All systems provide external control signals to trip external protective systems. All systems are self-resetting and are capable of withstanding and protecting against multiple events in rapid succession. These devices also protect against lightning or any other voltage surge that your system might encounter.

Here is a Bi-tron that's being built for a utility. The design here is for 30 kV and 10,000 amps, continuous duty, up to 1,000 degrees Fahrenheit without cooling. It's 36 inches long, 18 inches in diameter. It weighs almost 800 pounds.

This is a Bi-tron for 125 kV and 5 kA. It's 7 feet long and 2 feet in diameter.

For testing, this is a major move because we're talking about devices for which we haven't had test equipment to handle the kind of things that they're designed for. As part of our commitment to this area, we're constructing world-class EMP test facilities. I think most of you know that when you run electrical tests, the common practice is to test the high voltage at a very low amperage and then test the high amperage at a very low voltage, or you'll run a swept-narrow beam rather than an ultra-wide EMP pulse. Particularly for something new like our technology, but really for anything coming along, it's essential that devices meet realistic test conditions, under full voltage and full amperage and subjected to fields that may be there in real life.

AFS's EMP test facilities are capable of testing devices at up to 250 kV per meter and up to a million volts. We're not aware of anywhere else that kind of test is available.

This facility can inject simulated GIC fault signals with a user-definable waveform that accurately simulates real-world GIC waveforms up to 25 kV DC and 100 kA. That means we can specify what we think the pulse design looks like, but if you want to try something different, we can match the pulse you want. That's for GIC faults, and we can do the same for EMP. We'll be able to test devices at line voltages to 1.2 million volts AC or DC, under load conditions of 10 megawatts. And in sub-100 picosecond rise time, pulsed electric field environment greater than 250 kV per meter, and again, with a user-definable waveform that accurately simulates the real-world conditions.

While the facility is primarily for AFS production and research work, we'll make it available to other users, including competitors, because this is something that we all need to work together to protect the country. And while we think we have a good idea, maybe somebody has a better one.

The facility also can successfully create other EMP pulses as well.

This is our test cell number 1. You can see a person standing in the background next to the high voltage transformer, to give you an idea of the scale. These are 80 feet long, 40 feet wide and 20 feet high, with a door of 16 by 16, so we can get most things into this chamber.

In addition to the one we just talked about, we're building two more that are 135 feet by 50 feet by 50 feet. All cells will create arbitrary EMP waveforms and test at 250 kV per meter. All cells can duplicate magnetic field conditions of the largest transmission lines. They'll have digital instrumentation, greater than 25 gigasamples per second per channel on 12 simultaneous channels, and less than 17 picosecond resolution.

Just in case you're curious, here is what a 10 Hz, 15 kilojoule, 250 kV per meter pulse modulator looks like. In normal working position, it drops into the tank below it, and you can see a picture here on the right, already in the tank.

This is an artist's concept of a transformer protected from both EMP and GIC, with an EPS system. Each device senses the pulse coming down the line and connects to ground through a low impedance connection. The device implements field collapse protocol. Detection and operation are autonomous. The device provides hardened data output containing information on the system status and an EMP event alert. And it's available from about 4,000 volts to 1.2 million volts.

There are bulkhead-mounted devices that give transient suppression and EMP protection for shielded rooms.

AFS also has some interest in protecting from solar storms. Field collapse is our approach for both the EMP and GIC. We also make a neutral blocking switch for Kappenman method neutral blocking. It's available as an integrated system from Phoenix Electric or it's available to anybody that would have a need for that kind of a device. It looks like this; this is 35 kV, 100 kA, 25 inches long and 18 inches in diameter.

Just a word about what we're protecting against. If you're going to have a nuclear event, the fundamental physics governing the generation of the EMP pulse demands that if an E3 pulse is present, both an E1 and E2 will also be present. The reverse is not necessarily true. One can generate a non-nuclear EMP pulse that only has E1. So if you're designing a protective system, there is little sense in protecting just for E3 in an environment where you're expecting E1.

Just a final thought: we're a new company and we have new products, but we're not a garage operation. This is our building in Newtown, Connecticut, 250,000 square feet. It looks much prettier with the horizontal view, but this shows you the size and the magnitude of the facility.

So, thank you very much for your attention. I appreciate your interest.

MODERATOR (CHUCK MANTO): Let me introduce George Anderson, Chairman of Emprimus.

GEORGE ANDERSON: Good afternoon. One thing I've learned in about 40 years of being the engineer for safety and chief engineer for design is that often times, good thoughts don't accomplish very much for any purpose and results really depend on something that is built, tested, and put into service. This often requires a few years to prepare something like that, but real results only respect real functioning safeguards. In 2001, I heard of EMP, and that's now 12 years ago. It's kind of a slow pace that has me nervous about the whole thing. Another thing I think is important is that we have to have a little faster pace in the future. I think it's picking up this year—I see a lot of activity I didn't used to see: actual hardening of some sites, data centers and that kind of thing being reported around the United States in particular.

Gale was quite a find for me. In 2007 and 2008, we got together and founded Emprimus. He is both an experienced electrical engineer, and he has experience at Xcel—it used to be NSP (Northern States Power) in Minnesota—both as an engineer and as a lawyer for the company. He is experienced in high-level risk management and insurance—that was another stint he had. I hope you will listen to him carefully as he defines a few things we're trying to figure out about high priority things that actually have real products, and also some means of remediation of those risks.

And as you listen, please keep in mind a couple of things. Whatever you do, wherever you are, try to help advance the cause of protecting our infrastructure. And lastly, the words of an effective, impatient activist, General George S. Patton, I hope you'll ponder these words, "Fill the unforgiving minute with sixty seconds of distance run." I'll leave you with Gale.

GALE NORDLING: Thank you very much, Chuck, for inviting us and for all of your patience in being here. And thank each of you for advancing the ball in some way in this whole field to help protect our country.

One of the ways that we are trying to advance the ball is with some new products and extensive modeling in testing. So, with that, one of the first things we've heard from NERC (North American Electric Reliability Corporation) is that one of the biggest risks to the grid is if there is grid instability caused by a GMD (geomagnetic disturbance) event because of half-cycle saturation on the transformers, and it could be a blackout. But some utilities have said to us, well, a blackout will happen before there is any damage to equipment. So, let's take that as item number one.

Here are some recent blackouts in the United States. And you can see that the estimated cost from the U.S. government is in the billions of dollars to the customer. So, I think the cost to the customer has been overlooked in each of these situations.

Why is the blackout important for another reason? You don't just turn the switch to start all of these big power plants. They have huge feed water pumps, emission control systems, coal handling systems that take an enormous amount of power. So, you first have to start a bunch of other generators to even get the large generators going and then a coal plant that takes hours and hours for the boiler to heat up to be hot enough to even generate steam.

So, the 2003 Northeast blackout—obviously pretty real—here is what NERC said about that. Here is why we believe that procedures will be inadequate as a way to control or to protect the grid. There are many, many variables that will occur, whether it's the strength of the storm, the direction of the storm, what generators are in service, out of service, purchases, sales, on and on, time of the day, the load level, etc. So, utilities have said, all of their operators need specific simulation training for the conditions. And if the conditions are different, the operators won't know what to do.

Well, look what NERC said about the 2003 blackout—and it's essentially the exact same thing that they're going to say about a GMD event. Start from the bottom: "Inadequate training, inadequate anticipatory analysis, inadequate communication, inadequate voltage control." It will be all the same reasons. And guess what will happen? The blackout will last for a long time.

Here is a big item, and the utilities have not had the data perhaps to focus on this issue until now: Idaho National Labs tested for harmonic issues at the distribution level. The utilities have been focusing on the high-voltage level. At the distribution level, you get 33 percent harmonics at about 100-120 amps of current in the neutral.

Well, the IEEE (Institute of Electrical and Electronics Engineers) standard is 5 percent, 7.5 percent for a short period of time. That's going to cause damage to equipment, to motors, to rotors, to condensers, etc, etc.

Luis Marti from Hydro One came out with an IEEE paper that said, "You're going to have damage to rotors, generator rotors if you have more than 50 amps of current per phase." When we are designing equipment, we look at what the maximum credible threat is, in order to come up with a design standard. What are some of the maximum credible threats?

Data was filed for the Chester, Maine substation, thanks to what's going on in Maine. For 20 years, they recorded the GIC, the ground-induced current, in the neutral of transformers at that substation. John Kappenman correlated that data with magnetometer data from one of the nearest sources of a magnetometer. When you correlate that and put it together—and just look at the bottom line, because we've got to go quickly here—if you look at where there is 100 amps in the neutral, this position down here, you see that's at about 250 nanotesla per minute. So, if you ramp that up to 5,000 nanotesla per minute for a 100-year storm, you multiply that by 20. You're going to get 2,000 amps in the neutral.

Remember what we were talking about for harmonics at only 100 amps. And remember what Luis Marti just said, you'll have the image to rotors at 50 amps in the neutral. So, we're suggesting this is a pretty serious issue.

You've seen this before, the coronal mass ejection, the size of it compared to Jupiter, compared to the Earth. What does it do? Look what it does to transformers on the bottom two pictures. The left one was in South Africa, the right one was Salem, New Jersey, and you can see once again the size of some of that.

When we have done modeling, we've done extensive modeling with PowerWorld. PowerWorld was originated by a professor, Tom Overbye, from the University of Illinois. So, he has been able to simulate and put GIC currents on load flows to simulate what happens when there is a solar event while the grid is in operation.

And by the way, John Kappenman has some modeling, Siemens now does, GE now does, Mitsubishi, etc. This didn't exist a year or two ago.

Here are two individuals that have done an estimation of what the field will be for a 100-year event. At the top, John Kappenman's numbers are about 15.2 volts per kilometer, and there are 20 volts per kilometer on average for Antti Pulkkinen from Catholic University, although he has got a range of 10–50.

So, we simulated putting current blocking devices in the grid in Wisconsin, and we have a partnership that we're working with ATC (American Transmission Co.) in Wisconsin. They've given us actual transformer and line data and so forth, and so we did this. And you can see what happens the more blockers you put in. This is only 25. You can see the total GIC going down in the system and you can see the megaVAR (mega volt ampere reactive) loss going down.

For a while, utilities didn't have the software tools to say, "Well, I don't think the grid will even go down." Well, here we are, and you can see here that with no blocking devices, the Wisconsin grid will collapse at about the 100-year level for a solar storm. And then as we start adding blocking devices, you can see the grid becomes more and more resilient.

This is out of a total of 350 transformers, and this is blocking only 27. So, what we're saying is by blocking 10 percent, 15 percent, 20 percent, you get significant resiliency in the grid and it's not an overwhelming, huge, costly issue.

Here is another really interesting graph. Remember before, I said actual data could point to 2,000 amps of current in the neutral. So, we simulated with PowerWorld over in Wisconsin. We said what's the effect on the neighboring utilities if we blocked Wisconsin? Because we've heard from many utilities, "Oh, well, if we block, it will push it to the neighboring utilities, or vice versa, it doesn't do us any good to block until all of the neighboring utilities do."

Well, let's take a look at that. We'll call this a mythbuster. So, if you look here, this is before there is any blocking. We took the six highest interconnections from the Wisconsin grid to the surrounding states. But look how high the current is on some of these. Here it's almost 3,000 amps. So, under a simulation, it is confirming what the actual data could be. And then, of course, there is 2,000, 1,500, 100,000, etc. And when we put blocking in Wisconsin, look what happens—virtually, nothing happens.

Now, the reason it doesn't go down is we're not putting blocking in the neighboring states. But if you look at that, it completely debunks the myth of "we're going to add current to the surrounding states or vice versa." So there is no reason a state shouldn't go ahead immediately and do blocking. They will not adversely impact their neighbors.

This is the system that we started out with and is still offered for sale, which is basically, here you have the solid ground connection, and during GIC it goes through a resistor and a capacitor bank. That doesn't allow the DC current to go through, but the AC does. That's been tested. It's been simulated at the University of Manitoba. It's been tested at KEMA Labs in Philadelphia, and of course, Idaho

National Labs, that was talked about this morning. It was placed into a high-voltage grid and it operated just fine.

In fact, here is a kind of synopsis. These lines over here are with no blocking. So, when there is DC current injection of different levels, you can see that here, that's what the curve would look like. But as soon as block was put in and it was arbitrarily set to different time limits here, it immediately went to zero in every case where it was set and goes across. So, it just stops that altogether.

Our guys were out at the site, and when they were injecting DC, the transformer would be humming and shaking and noisy, and as soon as the blocking went in, it stopped just like that. So, we've come up with an even beefier system. When we did some testing and modeling, we found that in the Wisconsin grid, at the 100-year storm level, we checked the voltage level at every single substation, and with one exception, the highest that it was at was 2,000 volts.

The first model that we had was capable of operating at 4,000 volts. But there was one long line, 150 kilometer line. So, 150 times 20 volts is 3,000 volts. So, we said, for long lines in other states, maybe even for 30 volts per kilometer, let's make a beefier unit. So, we did that with two capacitor banks and it's now 8,000. So, it's 8 kV, which will handle, we think, virtually any situation. If there happens to be an extra-long line for EMP at 50 volts per kilometer, we can add more capacitor banks and it will still operate.

We also have a thyristor bank, which some utilities were saying, "Well, what happens if I have multiple faults during a GMD event?" Very unlikely. Very, very, very unlikely. But we said, "Okay, we'll deal with it." So, we put in a thyristor bank so this can fire repeatedly so that a fault will not destroy or take out the capacitors that are put there for protection.

So here is the upgraded model, and you can see that it's got two capacitor banks here at the top. The size of it hasn't changed: it's still 8 feet by 8 feet. We just pushed everything together closer.

We used SEL Electronics (Schweitzer Engineering Laboratories). We had electronics made by a California firm that worked just fine, but we said, "You know, for the confidence of the utility industry, we're going to use a proven Gold standard, if you will." So, we've entered into a further partnership with Schweitzer Electronics and, of course, ABB parts.

I'm not going to go into all of this—we don't have time for it—but we're demonstrating to the utilities that there is substantial backup and redundancy, with every component and so forth, so they don't ever have to worry about not having an effective ground as defined by the code.

A quick example to that is solid ground. The utilities, most of the time, want to operate in solid ground. But an effective ground under the code is still to operate through a power resistor and capacitor banks. This technology was developed by utilities, quite frankly, in 1983, 30 years ago, but they only had this. They didn't have the advancement of solid-state electronics, which now allows us to keep it in a solid ground 99 percent of the time, and then switch over to this only when it's needed. So we're trying to satisfy the issues that utilities have.

We did a study here as to any unintended consequences. If you get a bunch of these installed in the grid, is it somehow going to affect the operation of the grid? The University of Manitoba in short order said, "Nope, no problem." EPRI, the Electric Power Research Institute, has also just studied that and said, "No problem."

This is Schweitzer installing the unit over in the ATC grid in Wisconsin. The cost for this ranges—if you buy them one at a time—probably as high as $320,000, and the installation is $25,000–$50,000.

Now, when there is a solar storm warning, utilities—PJM, as an example, does it on the east coast, Con Ed, Dominion—they re-dispatch energy, meaning they back down base load generation and they start a bunch of peaking units so they have large generating capability. Instead of having two trucks running their engines at full speed in the parking lot, they want to back the trucks down and have 20 cars starting and running.

But the problem with that is that the cost of the energy is much higher and that cost is passed onto you. That cost, in one to two years, could pay for remediation of the whole grid, just to put this in perspective.

We also think that when transformers are humming and shaking and all of these things are going on, or you get 1,000 amps or 3,000 amps in the neutral, we think that blocking can certainly save the life or extend the life of a transformer. The average age of transformers today is 42 years. So, you put that in perspective, we're talking about an old fleet to begin with. There we are.

MODERATOR (CHUCK MANTO): One question.

AUDIENCE MEMBER: I'm Bill Harris, with the Foundation for Resilient Societies. My question for the panel altogether is—first for Bill Joyce—it has to do with the effect of—NERC has determined to eliminate a standard for physical security at even the critical electric facilities. However, the states have police powers, and we could have standards at the same level. So my question has to do with the intentional RF weapons, such as can be put in a van. Do you think, given the potential for this kind of damage at multiple sites with non-nuclear EMP, that there ought to be action at the state level, given that there is not going to be any action in the near future at the federal level?

WILLIAM JOYCE: I don't think doing a very good job of protecting your perimeter is going to help you at all, but I might ask Gale with his background, and a lot of other advantages, I know from a chemical plant, of having good protection. Gale could probably comment on the case of a power plant.

MODERATOR (CHUCK MANTO): I'm also wondering if the end of his question is: in the same way that the states have police powers, might the states be used to compel utilities at the local level to get the protection from either companies like yourselves, as an enforcement or a provocateur of protection that would create a demand for your product?

GALE NORDLING: Let me address this a little bit. In the recent weeks, the utilities have now had to focus on physical security based on the individual or individuals that attempted to shoot at the transformers for Southern California Edison. But different folks in the military have told us that there is an advantage to, and they are now looking at, facilities with a multilayered security approach, meaning you have one layer at the outer boundaries, and you change what you're doing for security at different levels as you're getting closer and closer to the heart of the operation. So distance is your friend, and you would want to put at least some kind of detectors—we also incidentally make those as well—but some kinds of detectors for various threats as far out in the perimeter as you can, with a smart type of video system, and then you want to do increasing protection until you finally would do the MIL-STD-188-125 and so forth on protection on the electronics themselves. So it's kind of a mixed answer, if you will. I think we're just learning about a lot of the security. I can ask you guys a question: how many of you, in all of your college days, had a single course, or a single professor, ever talk about security and mix it in with what was going on—whether it was a design standard or something else in engineering? I didn't have one—I don't know about you. So I think the world is changing.

MODERATOR (CHUCK MANTO): We have time for one more question.

AUDIENCE MEMBER: I'm Frank Gaffney with the Center for Security Policy. I just wanted to drill down for a moment on the finding of the University of Manitoba, and the industry's own internal research arc, that there isn't in fact some sort of bleed-out problem—I don't know the technical term for it—elsewhere in the grid if you protect part of it. That's not the industry's position, as I understand it. Could you explain how its own research arc is apparently at odds?

GALE NORDLING: Well, the studies that I talked about with EPRI and the University of Manitoba have to do with operational issues, like are you going to affect relaying in a location because of the insertion of blockers? But our studies that we just finished with PowerWorld would suggest you're not going to push the energy around on the grid to different systems, providing you've put enough blockers in your own system. It changes and settles down the whole level of GIC such that you are not going to adversely affect your neighbors. To our knowledge, this is the first study that's ever been done which concretely shows that.

Cost-Effective Financing Microgrids
for Critical Infrastructure

http://youtu.be/L1GqZKkru9g

Mr. Jeff Weiss, Cofounder and Managing Director, Distributed Sun

JEFF WEISS: Thanks everybody. So I guess I'm here a little bit for the comic relief. It's the middle of the afternoon. And we're going to talk about my favorite subject. Since I'm here, I'll talk all about me. My favorite subject is money. Now who knows where the money comes from? We're in Washington, D.C. Many people think it comes from the Capitol. Well, Congress is fresh out, I'm sorry. Many people think it comes from the utilities. Well utilities struggle to pay for things. And they have a business model where they're resistant to adding cost and adding structure.

So there's a fabulous place where money grows on trees. And it's called Wall Street. They have so much money there; you just have to know their rules, and you have to know how to go about it.

Structurally, we're going to talk for a little bit about solar and about microgrids. So, as we've been talking about EMP (electromagnetic pulse) and threat and threat protection, one approach is to have resilient microgrids. And, of course, resilient microgrids can be at the level of a hospital, at the level of a community, at the level of a piece of critical infrastructure…but of course they have to be funded. So that's where the key comes in.

The interesting thing about the world today is there is a nice analogue, which is what's happened in the distributed solar industry. That's really where I'm going to lead and what I'm going to spend some time talking about—to explain money, to explain risk, to explain finance, to explain how to extract money, and how to fund all this.

From an EMP point of view, though, I wanted to level set and to say that interestingly, almost every microgrid that gets deployed or that is talked about being deployed includes solar as one of the constituent parts. So a microgrid simply is a grid that's small instead of big. That's what the word "micro" means. It's usually contained. It's for a facility. It's for a community. It's for a smaller area. It could be for a military installation.

It always has two or more forms of power. So a generator on your facility is not a microgrid. A generator is a generator. But a generator is part of a microgrid. It's one of the constituent parts. So there are always two or more forms of power. Microgrids always have smart grid technology, quite simply because, if you've got multiple forms of power coming in, you want to have a way to control it and to push it back and forth.

Microgrids always want to have a black start capability. So again, coming back to EMP and the threats we're talking about, you want to have the ability to turn it on when other things go off. And, of course, then, after that, there are other forms of requirements that really depend on the need or the locality. The big and expensive one that's quite often talked about and added is cybersecurity.

So, of course, the power can come from solar; it can come from CHP (Combined Heat and Power); it can come from fuel cells; it can come from hydropower, wind, geothermal. The lovely thing about power is, you just have to look at the location you're in and determine what's available. But you always want to have at least one—if not two or three—renewable forms of power. Because quite obviously, while you're talking about something that results from an EMP or just a long-term grid outage, if there's no infrastructure to get something from point A to point B, you need to be able to generate it right in situ.

So I've got some fun words up here: avoid, reduce, accept, or transfer risk. Solar is way ahead in the development of itself, from an industry structure point of view, from microgrids, which are much earlier-stage. They're less defined. Solar is kind of a widget and it's getting deployed very, very broadly now. What most people know is that the hard costs of solar have come way down—95 percent of all solar in the United States has been deployed since 2008. A big part of the reason is that, before then, it costs $5–$10 a watt to deploy solar. Now it costs $1.50–$2 per watt. Those are called the hard costs. The biggest elements of those are the panels, the inverters, the electric cost, the labor. It's the stuff that's fairly tangible and easy to see.

What's behind the curtain, and the really hard problem, which I'm going to spend the next 10 minutes talking about, are soft costs. So soft costs are those that you don't actually directly buy, but everyone pays for. So soft costs are everything to do with the cost of capital, the cost of unforeseen risks, the cost to close a transaction, the cost to market and attract everything you have to do to assemble all the parts and all the constituent customer parts. So soft costs are very high.

Until this morning, I just would have told you that they're high, and you would have had to agree with me. But now I can point you to an NREL study that's just come out: the National Renewable Energy Labs have produced a 75-page document that documents soft costs. And they've now documented that it's the last frontier, and it's the most important problem to solve.

So, from an engineering point of view, what that means is, we've gotten all the hard costs into a zone where they're pretty understandable and they're now cost-effective. But we have not done that with soft costs, which is why it's now Wall Street's turn. Soft costs ultimately have to do with cost of capital. So if I go to my friend Bill over here, and I say, "Bill, have I got a deal for you. Here are 20 projects that I think are right up your alley. Would you please invest in them?" He'd say, "Sure, I like you, Jeff, you seem to be a good guy. I'll look at them. I'll go to my partners at J.P. Morgan; I'll have their analysts look at them, and we'll get back to you in a month or two." And that's a reasonably normal banking-like business process.

The problem is, in the world of solar, which is very far advanced—and then certainly in the world of microgrids, which is much less advanced from a unit price package structure point of view—the problem is that the diligence process that's undertaken at the institutional investor level is very opaque, and is different for all of them. J.P. Morgan, Morgan Stanley and General Electric all do it differently. There isn't a standard process. No one asks the same questions. Therefore, the cost of interacting, of completing the diligence, and of knowing what you're doing is actually extremely high. And that opaque process and high cost is causing the preponderance of projects that get initiated to fail.

In the country, 95 percent of projects that are initiated fail. It doesn't mean they fail from an electricity generation point of view. It means they fail because they don't get money. They don't actually get built or deployed.

So how, in a more business-like structure, does one get around this? You do it with standards. So again, going back to my bringing Bill these 20 projects, Bill would be better off if, instead of just my saying to him with a smile on my face, "Here are 20 excellent projects. I think you'd like them, don't you?" and let him nod his head, it'd be better for Bill and for his friends at J.P. Morgan if he was viewing them in a standard process, if he was viewing in an industry that had a standard structure so he knew exactly what it meant, and if those standards were driven by industry, because industry is what matters, and that there was a credible path to broad adoption.

That's what I'm going to talk about. This is an introduction to truSolar, which, in the solar world is the standard process that industry is developing, and my firm is leading, to create an approach to diligence and risk mitigation which is replicable and broadly adopted. And it's absolutely almost cookie-cutter applicable beyond solar to each individual form of renewable energy, as well as to microgrids, which is our business intention.

The next issue is standards; we've asked a lot of people about this. Eighty-three percent of developers—people that are out there in the solar world—say that their business models are misaligned to financing terms and to pricing. Well, that's uncomfortable. Who likes to start on something where you feel that 83 percent of the people who are giving money don't even speak your language or don't understand what you're doing? Ninety percent of originators report more than 50 percent deal dropout. Well, talk about a high cost. The cost of that which you start to work on and just work and work, and send your people out, and maybe spend some money on, that don't even get done…that's a very, very high cost. So we need an IT-enabled deal flow to help structure this and make it a transparent and fluid process.

This system we're working on is driven by industry. We had the good fortune in starting it to have Standard and Poor's and DuPont as our first partners, ABB as a partner, Booz Allen, I can go on. They're all publicly listed, but we've got seven publicly traded companies, $100 billion of revenue among them, and another dozen private companies, all of whom are significant in the value chain. They represent some of the largest manufacturers and the largest service companies and largest insurers.

I'll give you one example of where this is going from insurance. From a cost point of view, the largest single expense, if you own and operate a solar facility—after you commission it—is property and casualty insurance. Why? It's because this is still an adolescent cottage industry. So in America, you think microgrids are early-stage; solar is less early-stage but still adolescent. As an adolescent industry, you go to the property and casualty industry, as we have, comprehensively, and you say, "Why do you charge so much for property and casualty insurance?" They say, "Well, we don't have enough data. We don't have enough experience with losses because there haven't been enough around for enough decades."

So, what they do is, it's not that they make it up. They charge enough to be sure that they're protected and covered. So, when you go to the insurance industry and say, "Well, we're about to have a rated process where, instead of giving Chuck a portfolio of more generic projects, you now have a portfolio of rated projects, kind of like a FICO score for solar." They say, "Well golly, we're going to charge less for property and casualty, and we'll have a truSolar-rated property and casualty product." That will come out in the next year.

I guess I've talked a little bit about this. That's an example of adoption. The way this is going to be adopted is this process, and most people probably know the word FICO score from consumer finance. This is a commercial industrial analogue to that—it's more complicated, but it's not unlike a FICO score for commercial industrial solar projects.

Publishing it will be complex but it will be published. So it'll be like putting the tax code online. We can all go online, and we can pay our taxes by downloading forms at IRS.gov and the rules and pay them legally. Or, we can also go to software like TurboTax and choose to use a forms-driven process. Both will exist in the near future.

So where is this leading? These points, I believe, are totally analogous to the entire conversation for today. I'm the non-technology comic relief part of the afternoon. This is all about lowering the cost of capital. Money costs a lot of money, so in the world of commercial industrial solar projects, the number right now is nine percent. Now that doesn't necessarily mean anything out of school, but to get a Wall Street investor to invest in Bill's solar project, he needs to show them that he can give them a yield, which is nine percent, after tax, unlevered.

Well I will tell you that, in the world of things that are basically bond-like returns, nine percent is a pretty high interest rate—it's a very high interest rate. Why is it such a high interest rate? Because it's still a cottage industry, and because there isn't truSolar, and there isn't a rating system. So the thought is, when we say lower the cost of capital, Bill also used the word log scale before in showing his graphs. If we as industry get the cost of capital to go from nine percent to eight percent, that's logarithmic growth in the number of solar projects, and therefore, the number of microgrid projects which will be invested in, in America. If we get it from eight to seven percent, it just goes beyond. So this is actually very exciting.

And the final thing is: It enables securitization. The Holy Grail on Wall Street, the place where the money grows on trees, is securitization. Wall Street loves to invest hundreds of millions of dollars or billions of dollars in new things. That's how the mortgage markets took off; that's how the derivatives markets take off. There is goodness on Wall Street. There are trillions of dollars to finance anything— including microgrids and commercial industrial solar—when they can sell securities. They can only sell securities if they can underwrite the securities—that's what the SEC is for. There are laws that say you cannot sell security unless there is an underwriting standard.

This is all being set up to allow securities. Now let me give you some data, because 2013 is the first year in history that there have been securities sold in renewable energy. This is very exciting. So I've been around this for a number of years, now. And everyone has been saying, "Wow, this will all be good when we have securities sold, when Wall Street, those guys up in New York sell securities." But there haven't been any.

So finally, now, there are. SolarCity, which is one of the most successful solar companies, period—residential solar—sold a bond three weeks ago. The bond is a $54,500,000 bond. They got Standard and Poor's to rate it. It's a triple B-plus rating. And the yield that came out on their bond is 4.8 percent. Well, that's a lot less than nine percent. So we're very happy about that.

NRG, the large utility, sold a bond company, which is called a yield code, which went public last summer on the New York Stock Exchange. They put $2.2 billion of assets into it. When it went public at $23 a share, it was yielding 4.5 percent. It's gone up to $37, so it's now yielding 2.7 percent. Again, yay, 2.7 percent is a lot less.

I have one more thing I want to do. This video is hopefully going to give you just a quick mid-afternoon bit of humor about where truSolar is going. And then I'd be happy to answer any questions.

[Video]

MODERATOR (CHUCK MANTO): Great, thank you. As the next panel comes up, do we have a question from the audience?

AUDIENCE MEMBER: I'm Terry Hill with the Passive House Institute. How about a city block of very energy-efficient houses, all with solar and a microgrid—could we finance that?

JEFF WEISS: Great idea. The answer is, anything is financeable if one understands the contracts and the risk. It needs to have, in today's market, over a nine percent after-tax unlevered yield.

MODERATOR (CHUCK MANTO): So the answer is of course. And we're ready to do one almost right away, with a little bit of homework.

Cost-Effective EMP Protection for Communications Networks and Power Sources

http://youtu.be/VsSdLPqAbkw

Dr. George Baker, Professor Emeritus, James Madison University
Mr. David Oppenheimer, Pathion

MODERATOR (CHUCK MANTO): Dr. George Baker is Professor Emeritus at James Madison University. Dr. Baker was involved with something called the Defense Threat Reduction Agency, formerly called the Defense Nuclear Agency.

David Oppenheimer is also here and is going to take few minutes to look at what you can do to mix solid-state energy storage with energy generation. His company is responsible for a piece of breakthrough technology on the energy storage side to make it even easier to get guys like Jeff Weiss to help you finance your solar package.

I'd like to start by turning the floor over to Dr. George Baker.

GEORGE BAKER: Thanks, Chuck, for inviting me. I'm delighted to be here. I don't whether this is an exercise, this late in the day on Friday afternoon, on my ability to entertain you or your ability to withstand boring facts. But we'll press on here.

So, I think Chuck did a good job. I have worked for the Army, protecting tactical systems, that's where I started, against EMP (electromagnetic pulse). Then I transitioned into the Defense Nuclear Agency, looking at more strategic systems and developing standards. And then finally, with the Defense Threat Reduction Agency, I was the Director of their hardened target research facility. Then I left there and I just retired from James Madison University—I was Professor of Applied Science there. So, I feel like I've been through the gamut, but hopefully my experience will be of some value to the quest here.

The agency that has the most experience or the department that has had the most experience with EMP protection is the Department of Defense. We all stand to learn a lot from their experience, their hard-learned lessons. So most of what I want to say today is based upon my experience developing EMP protection techniques and standards at the Department of Defense.

I want to start with some quotes that I think are particularly appropriate when we think about EMP and what's been done and what hasn't been done. I'm a UVA graduate, so I have to have the obligatory Thomas Jefferson quote, "We're not afraid to pursue the truth no matter where the search may lead." I think this is a very important adage vis-à-vis EMP and GMD (geomagnetic disturbance) effects.

And then this Henry Kissinger quote I like—and this is what's happening: "We're avoiding coming to grips with problems by making projections of probable events which require little action." That's what's happening. We are making "assumptions about the course of history that are serving as a substitute for effort." That is happening.

And we need to remember, but especially the policymakers need to remember that they're responsible not only for the best thing that could happen, the best-case scenario, but also for the worst. We can't entrust the survival of our society entirely to guesses, and I'm sorry to say, that's what happening.

There is a lot of work to be done. We need to get with it, and hopefully, what I have to tell you based on the DoD experience will help.

First, just some ways to think about cost. I hear a lot of people talking about cost, but there are some very, very, very simple ways to think about cost. EMP is one of these high incident, low frequency, or HILF, effects—actually a better acronym is high effect, low likelihood; these are HELL effects.

The cost for this kind of protection is offset by normally unobserved benefits. If you don't an EMP, you can't tell. It's like snapping your fingers to keep the elephants away. Anytime I talk to somebody about EMP, the first thing I have to do is do an existence proof, a credulity proof, and that gets to be very old—and I think that's changing. I hope it is.

People say, "Well, this is never going to happen, and if it were, there is no way I can predict the probability." Well, that's wrong. There is a guy named Martin Hellman, a Beta Tau of the IEEE Mensa Society, who has done very elaborate predictions of the probability of nuclear attacks, and he has bounded the probabilities, and they're very high that we will have a nuclear weapon detonated in an unfriendly manner. So, Google Martin Hellman.

The other thing is EMP protection is not highly visible, and when it works it's often taken for granted by management and customers. And we find, in DoD, it's very easy to forego the maintenance of EMP protection because of this lack of visibility. Corporations make their money by selling products and services, but protective security with which they operate is often an afterthought at best, and these HILF- or HELL-type effects are the lowest of the low in terms of the afterthoughts. But the problem is—and this is where we're going to go into this very simple cost modeling—is that neglect of EMP protection has very serious financial—even existential—consequences.

Obviously, we need to remember that businesses know more about themselves and their systems than we do, and also, they know more about potential threats, and that's where we need to educate and create awareness. The benefit of protection systems is not easy to characterize, but it can be done, and I'll try to show you how here.

We need to treat the problem by considering losses and using combinations of failure probabilities and associated consequences. The basic idea is that more protection means lower risk of losses. Businesses understand losses. So you want to do the cost arguments in terms of lowering the risk of losses.

The other thing is we have to remember that perfect protection is not possible. We need to recognize that. You're never going to have a perfectly survivable system, so quit trying—that was a message that came through pretty loud and clear in several other presentations today. But we have to begin thinking about cost here. So, here is a line of increasing protection until you get to the point of perfect protection. But you're never going to get there. And then in the opposite direction is the direction of increasing risks. The less the protection, the bigger the risk.

So, we can draw a protection cost profile: the smaller the amount of protection, the smaller the cost. And the cost will go up in some asymptotic way—you're never going to get to this perfect protection point. You can waste a lot of money on protection, and we want to avoid that. But this line is not adequate. The protection profile doesn't allow determining the optimum protection level. That's the basic message of that chart.

There is another line we can draw, and that's the loss profile. If we don't protect, the cost can be very, very high—in fact, this Lloyd's of London report was talking in terms of single-digit trillions of dollars, if we were to suffer a major grid collapse. Now, if we're thinking about communication systems like FirstNet, you're going to need that communication system to restore the grid. So, the cost of not having the communication, and the ground awareness that you get from a communication system, I think also you could put the cost up in the trillions of dollars for not being able to restore the grid in a timely fashion. So, both losing the communication system and the grid communication engender similar costs.

If we increase the protection, that brings down the cost. Of course, you will never get to perfect protection. This [declining line] is the value of the losses as we increase the protection.

The bottom line is: the greater the potential EMP damage is, the bigger the benefit of preventing it. And as I say, for long-term outages, you're talking about things that are on the scale of gross national product.

If we superimpose these charts, we start to get an idea that there might be some optimum cost where the amount of money we've invested in the protection equals the amount of losses that we've prevented; and any costs we occur that are beyond this point are probably wasted.

So, you have this Point E, where the protection expenditures are justified. And to the left of E, there is probably some benefit. If you can keep the cost down here, it's even better, but here is the optimum cost. We have to have some idea of how much the losses are so we have some basis for justifying our investment.

Now, if you look at the real situation for EMP, the protection profile, the cost of protection is really low, compared to the value of the losses. I could have plotted this in a log plot. The protection costs are very low. In fact, they're so low that you probably can get pretty close to a really effective—not perfect, but a very effective protection for a very low investment. So that's the basic way of the thinking about cost. It's very simple, but I think it helps us as we go forward.

To gain insights into the interplay between risk and protection on the horizontal axis, let's look at the evolution of EMP protection philosophy within the Department of Defense. In the Department of Defense, we had a very, very intense debate from the early 1960s to the early 1980s, and it was so intense that there were people who lost their jobs because of the conflict between these two. And the basic conflict was caused by this tailored hardening approach, where you're protecting at the box level. You go into a data center and you say, "I think this box might vulnerable because it's sitting over here near the door; and these boxes might not be vulnerable. This other box, since it's where the power line comes in, might be vulnerable." And so you're doing a box-level protection. There are some places in Washington here where this was done, and it was almost like people were hanging surge arresters on lines sort of randomly. It had much lower initial cost, but it was based on a lot of guesswork.

Then you had people who thought, "No, you just need to put all of your electronics inside a shield." That's going to cost more, but at least you'll have a much better idea of what you've bought in the end. So in tailored hardening, the emphasis was on surge protection—voltage arresters, that kind of thing—and in global hardening, the emphasis was on shielding. The good thing about the global shielding was that it was much easier to do testing, and it made it easier to maintain the hardness [of the system] over the life cycle. Note that there was no attention to operational procedures. Unfortunately, we have people who think you can harden systems by just developing workaround procedures. In the DoD, we never thought about that because we were protecting systems that were time-urgent, systems that had to be able to respond to a missile attack, where you only had 30 minutes of warning.

Now, think about the power grid problem. How much warning are we going to get for the solar storm hitting the power grid? Anybody remember? About 30 minutes. So the amount of warning that we're going to have for power grid is exactly the same as the amount of warning we would get from a missile attack. We're not going to have time to do operational procedures. That's another reason not to do operational procedures.

As I said, there were people who lost their jobs because the conflict over these, and the debate precipitated two high-level reviews, one by the Defense Science Board, the other by the National Academy of Sciences. We know it's very difficult to predict EMP effects. I'm not going to dwell on this—it's going to take too much time to explain—but the idea is you can predict the currents and voltages outside a system, but as you go down inside the system, it's very, very difficult to predict what kind of currents and voltages you would have at individual components. And this is where the tailored hardening was trying to protect. That's really the crux of the issue.

So, what you do is you keep the energy out of the system, and then you don't have to worry about predicting what happens inside it. That's the idea: you put up a shield barrier. The National Research Council of the National Academy of Sciences said that they were very concerned about prediction uncertainty, as I mentioned, and they said the soundest results can be obtained where the stress within the system is controlled through integral shielding. The great emphasis on developing better and cheaper makes for virtually complete and effective shielding of systems. That was the idea.

They also said that we needed to develop design strategies that were testable, and if the system is shielded so well that virtually nothing gets through, it makes the testing much easier. And then they said doing the tailored hardening approach—the component-level protection rather than the system-level—carries much more risk of vulnerability. And again, they pointed to the inaccuracy of the analytical predictions.

Then the Defense Science Board came to the same conclusions: "tailored design...too much analysis... hardening components...the component level is very difficult...Best approach is a minimum number of contiguous shields." These two groups came to virtually the same conclusions.

So the DoD has used this system-level global shielding approach, and the idea is to make things that are electromagnetically simple. We know how to do that. The engineering is very straightforward. Also, you need to be able to maintain the hardness—you can harden a system and not maintain it and surveil it. It's like throwing your money out of the window.

And this is not all-inclusive, but here are some systems that the DoD has effectively hardened using the global shielding approach. And we actually developed a military standard (188-125), completed in 1990. And we were able to get a tri-service group together, with Army, Navy and Air Force. We are still talking about the major issue of "box- or system-level?" but the standard requires protection at the system level, and it applies to all DoD systems. It specifies protection measures, not vendors. I talk to people, I say, "We need standards for protecting the power grid." They say, "No, all you're trying to do is just get your vendor buddies funded." But the standards, if properly written, say nothing about vendors, and the DoD standards say nothing—they just have the protection measures: the decibels of shielding, the voltage residual levels and current residual levels inside that shield. You can do it however you want, as long as the voltages are limited when we test it.

Here is a notional schematic of how you do it. So, you have the shield barrier and every penetration is protected. Every conducting penetration and breach-type penetration—the doors and the pipes—is protected.

So you have a quiet zone inside here, and you can put anything you want inside here. You can put commercial, off-the-shelf systems, you can put any kind of electronics, because inside this shield is, electromagnetically, very quiet.

And I should mention, these neutral blocking devices are exactly the same idea as a barrier. If you use that capacitor blocking device, you can put anything you want on the other side of it. And it also takes away all the uncertainty of what the highest GMD current is that you can have. GMD currents can vary over a factor of 1,000 and it wouldn't matter, a capacitor is going to block it. On the receiving end, you don't have to test all the transformers, because if there is no DC current coming through, it doesn't matter what you put on the other side of that capacitor.

Here is a picture of the defense satellite communication system ground terminals. We hardened those using the MIL-STD-188-125. And there are some ways to reduce cost: you don't have to shield entire buildings, you can isolate your critical electronics inside shielded rooms. There are lots of people who build these. Or in some cases you can just put them inside shielded cabinets that you install with fiber optic interconnects.

The other thing I should mention is that Bron Cikotas is doing a lot of good work on looking at different ways of dielectrically isolating electronics from the outside world. I encourage you to talk to him about his results.

Just by way of summary, protection costs are well justified. If you looked at the cost of the consequences, the costs [to harden] are a small fraction of system cost for new systems. Our experience in DoD is 2 to 5 percent of the cost if you come in at the beginning. The costs are manageable for protection of existing systems, especially if you can isolate the critical electronics. EMP protection has a lot of other benefits, in terms of protection against electric power outages, lightning, general power line transient, GMD, electromagnetic interference, signal emanation security (a thing called TEMPEST) and RF (radio frequency) weapons effect. The EMP protection helps you in a lot of other ways.

And I have to emphasize the importance of protecting the backup power. That has to be integral to your approach. We've done this—the Department of Defense has hardened, I think it's fair to say hundreds of systems using this approach, so it's doable. That's all I have.

MODERATOR (CHUCK MANTO): Thank you very much, George. If you work in the data center world, which we'll be talking about in a little while, there are a lot people who spend a lot of money just trying to keep equipment protected from what they would call "dirty power" and things like that, which degrade systems and cost money. So, what I'd like to do is have some questions in a moment. But before we do, I'd like to have David Oppenheimer from Pathion come up and get us interested in what's coming down the pike a little bit, in solid-state storage of energy.

DAVID OPPENHEIMER: Thank you for having me here, Chuck. I appreciate the opportunity. I've worked with Paul Rich and his group here at the PSO (Policy Studies Organization) for the last few years and always enjoy coming out.

I'm with Pathion, Inc. We are an energy company, with headquarters in Los Gatos, California. Fundamentally, what we are is a technology company—material sciences. Those sciences came out of groups like Los Alamos National Labs and Berkeley National Labs.

What has happened in the energy technology and particularly storage of energy in the last 10 years is a revolution on the basis that how batteries go together has changed. We're familiar with the physics of traditional batteries, a little chemistry here, a little chemistry there. The answer is that a little of that is not quite so correct anymore. And that revolution in knowledge is going to cause a generation of batteries that will be significantly better than the kind we have today. That being said, what is the use of a battery in this solution set? How do we make microgrids that make sense? Some of this is just plain strategy.

Let's say this building has multiple power sources. It has solar, it has natural gas, it has generators, and it has power coming in from outside sources. But one of the best techniques you can use is take all of that power and pump it through your batteries. Why? Because all of the issues that we've talking about here don't pass through those batteries. The batteries give the building clean, even, uninterruptable power.

Let's go a little bit further. We have peak usage periods that have dogged this country, frankly, for a long, long time. We all get the bill, don't we, that SmartMeter—that's not our friend, is it? So, we know that we're paying for peak power at a real premium. What if this microgrid was under the control, partially, of the local utility? And so I could shift the building purely off the grid three or four hours a day at the right time. Suddenly, I even out my power usage around the country. There is a real economic value to that, isn't there? That actually changes the proposition here.

What we have to do to make this solution work to avoid EMP vulnerability is upgrade this network in a way that takes care of that and creates financial value to do it. What we're offering forward is the capacity to have batteries that are deep cycle, that can take a building on and take it off the grid. And whether that is in a pulsed manner, so that building is on for 10 minutes and off for 10 minutes, or on for a day and off for a day, these are all strategy issues of how you deploy the technology.

But fundamentally, you can isolate a building, and by doing that you provide clean power, you provide a barrier for EMP, and you create economic value by removing peak usage periods. You even those out across large areas. That's an alternative to how microgrids might be viewed in the future.

When I was in nuclear nonproliferation, we were reminded that there were three things that you always had to do. The first was: tell somebody they have a problem. Once you got a consensus that they understood they had a problem, you went onto a technology that could solve that problem. And lastly, you had to find a way to finance that solution, and that's the kicker most of the time. So, the answer here is if you can't find actual economic value in improving the network to do this, it's going to a very difficult measure. But I believe there is economic value in upgrading the network this way. And it's not about the utility owners, it's about the power users. Thank you. I appreciate my opportunity to speak.

MODERATOR (CHUCK MANTO): Thank you. I have a quick question myself for you. So, it sounds like a great set of strategies. You have a sense of where we are today with the cost of, say, energy storage. Let's say that all the energy storage in the country was costing us one dollar. With the new technology that you're seeing, some of yours, some of others, a few years down the road, what will that new technology do to that cost? If all energy storage today costs one dollar, how much will that same amount of energy storage cost when we realize the benefits of these new technologies?

DAVID OPPENHEIMER: Well, I think the question is the transition from one modality to another modality. Batteries have a cost, and that cost is never going to go to zero. But because they're going to have much, much longer life, and be able to perform effectively over a much longer time and depth—they can provide better energy output, they can provide power output. There is a difference between power and energy in batteries. That cost is simply neutralized or managed by the savings that you have, similar to solar systems.

MODERATOR (CHUCK MANTO): So, it's going to get better?

DAVID OPPENHEIMER: It's going to get better.

MODERATOR (CHUCK MANTO): By a teeny, weeny little bit or by a meaningful amount?

DAVID OPPENHEIMER: Well, if we were to just to look at battery and the battery were 50 percent or 100 percent more powerful or more effective than a current battery, then by definition, it's half the price.

MODERATOR (CHUCK MANTO): Wow, half the price. It sounds like we're almost there. Very good. Now I'd like to give the audience a chance for a question or two.

AUDIENCE MEMBER: I'm Bill Joyce with AFS (Advanced Fusion Systems). I'm in the dark ages on batteries, but I know when I try to charge a battery, if I put too many amps through, the battery is toast. These batteries right now can take this square-wave EMP pulse coming through with all of that energy and they're unaffected, unlike solid-state devices?

DAVID OPPENHEIMER: Well, there are larger systems. You would be able to protect against a system deployment in the way that has been described by this group. So, you would have filters in the front end and you would take the power into the batteries. And yes, they're much more resilient than the systems that would normally be taking that power directly.

BILL JOYCE: Okay. So what you're saying is there is a system in front of the battery to screen out EMP?

DAVID OPPENHEIMER: That's correct.

MODERATOR (CHUCK MANTO): Great question. Thank you, Dr. Joyce. I see a question here and then back to you.

AUDIENCE MEMBER: This is Bron Cikotas. I'm aware that West Virginia in some cases is using battery banks to take care of their peak power load capacity requirements, rather than building new power line capacity to bring in the power from somewhere else. Can you tell us a little bit about that? The other thing is if you look at, for example, the Nissan Leaf battery, that battery has the capacity of about 20 kilowatt hours. Its cost is somewhere in the area of $11,000. And they're now configuring those batteries not for the car use only, but for whatever use you may want to put it to.

DAVID OPPENHEIMER: Being a West Coaster, we hear nothing of Virginia's projects, I'm afraid. So, I can't comment as to how they're doing that. But that does speak to the issue of evening out the cost of the peak. The peak is so significant that even at that cost model, you're saving money fairly quickly.

Now, when it comes to the larger cells and the cost per cell, it is a case-by-case basis of how you deploy it. And I really can't comment directly to that.

MODERATOR (CHUCK MANTO): All right. Two more quick questions.

AUDIENCE MEMBER: Paul Kainen, from Georgetown University. What is the status of legislation on providing safe power for cities, if there were particularly a geomagnetic event of some sort that knocked out power to cities? You'd be talking about millions of people in jeopardy, especially if the power couldn't be restored for months. How much good would batteries do to provide safety—and more specifically, what is the status, not so much of national defense protection, but of public health protection in terms of power vulnerability?

GEORGE BAKER: The status of legislation is stalled. We lost the GRID (Grid Reliability and Infrastructure Defense) Act, that was stalled, and now the SHIELD (Secure High-voltage Infrastructure for Electricity from Lethal Damage) Act, I think, is following suit. This legislation would empower FERC, the Federal Energy Regulatory Commission, to issue or to mandate standard protection of the grid. So, it's going to be really important to pass that or similar legislation. There is this new SEPA (State Environmental Policy Act), which I think also would help in that regard to be able to ensure that the grid is protected. But regarding the battery question, I'll pass the microphone.

DAVID OPPENHEIMER: Please restate your question on batteries?

PAUL KAINEN: How much will batteries help if you're without power for weeks or months? The battery supply is for days, right?

DAVID OPPENHEIMER: That's correct. Actually, it all has to be part of an integrated solution, and that is that you need off-grid power creation. It evens out these situations so that you are not vulnerable and losing systems, but it doesn't create the power.

Now, it does go a lot longer than batteries that we conventionally understand today, but you still have to have energy source.

PAUL KAINEN: It could be very helpful then—the most important thing would be to coordinate rescue efforts, which would be enormous, and you'd need to have some power to provide the communications.

DAVID OPPENHEIMER: That's correct.

PAUL KAINEN: Even if there wasn't enough to heat houses and so on.

DAVID OPPENHEIMER: It is.

MODERATOR (CHUCK MANTO): And that Act he was referring to a moment ago is the Act that would be a Homeland Security-oriented planning activity. That looks like it has a pretty good chance of passage, from what we've learned. One more question.

AUDIENCE MEMBER: I'm Alan Roth with AFS. I want to refer back to something that Bill Joyce said earlier about our look at MIL-STD-188-125a and we see this as weak. And George mentioned the shielding that's been studied and shown to work over many, many years, that the military uses. But we have a special circumstance, I think, with our non-nuclear EMP, our EMP generator that has shown that the 188-125a failed at least with the testing that we did. And what I'm saying though is we're not saying, "Well, believe us" or "believe him," regarding shielding. We're planning to do testing, and we have really remarkable test facilities to do this, to bring onboard George and others to the realities that we think are there, that they will see through our testing. And I would say just to stay tuned-in to see what happens, because shielding really could be an issue that you may not see now that would be in the future regarding this.

MODERATOR (CHUCK MANTO): I think we'll give the panel or George a chance to answer that. I think too, it's like what you're basically saying, Alan, is that we're looking at a different threat than they were looking at. They've looked at one threat and you're trying to say well, there is another type of threat. George, did you want to comment on any of that?

GEORGE BAKER: This is important, and that's a very good question. The EMP standard protects systems from 1 GHz and below. And the sweet spot for the RF weapon, the non-nuclear weapons, spans the frequency range from 10 MHz up to 10 GHz. So there is a frequency band from 1 GHz to 10 GHz, where the EMP protection is not designed. Now, that's not to say there might be some inherent residual protection in that regime, but if you're worried about the RF weapons, the non-nuclear weapons, yes, you need to be a little more fastidious with your shielding and your door seams and your wave guides.

MODERATOR (CHUCK MANTO): They mentioned the opposite kind of a problem. EMP issues are when an electromagnetic field is impacting you. And the opposite also happens when you are typing on your computer in the old days, you emanate little radio signals and smart guys in a van across the street can capture those with an antenna and then decode everything going on in your computer. So, that program was called TEMPEST, which is almost the opposite of EMP. You're shielding it so things don't get out.

So, what's interesting is to blend the combination of those problems so that you're looking at all of the issues for things coming in that you don't want, as well as things going out that you don't want. So, harmonizing those would be very interesting. A number of us have been talking about harmonizing all of the MIL specs and all of the problems into sort of a harmonized standard for all of these. And we only have conversation and emerging industry practice about that. So, all of these are ongoing issues.

Energy Security Impacts on the Data Center Industry

http://youtu.be/F7tetRLcApc

Mr. Thomas Popik, Chairman, Foundation for Resilient Societies
Mr. Michael Caruso, ETS-Lindgren
Dr. George Baker, Professor Emeritus, James Madison University

MODERATOR (CHUCK MANTO): You're going to have a very interesting panel talking about something that sometimes people overlook, which is very critical—called data centers. Data centers use a significant amount of power, and they are expected to use a lot more but they are highly vulnerable also. They are going to talk about the importance of them, their vulnerability, what we might do to mitigate it. You have top experts who are going to be doing that now.

And what I'm going to do is introduce Tom Popik, who is the Chairman for the Foundation for Resilient Societies—he spoke to us last year. His comments are in this book that you can pick up at the back. And you can also "YouTube" all the participants from last year. Before too long you will be able to do that again with all the participants this year. And so he is going to explain a little bit about himself in brief, lead the panel and introduce his other panel members. Tom Popik.

TOM POPIK: Thank you very much, Chuck. Again, my name is Tom Popik. I'm Chairman of the Foundation for Resilient Societies. We're a nonprofit group that is concerned with protection of critical infrastructure. There are several of us who have been here at this conference. We're going to be talking today about energy security impacts on the data center industry. I would also like to introduce my fellow panelists: Dr. George Baker, who is also a member of our foundation, and Mike Caruso.

So our agenda, briefly is—I'm going to talk about some things that we learned from going to a data center conference. Now, in order to get EMP (electromagnetic pulse) protection for the grid, we're going to need some industrial constituency. We need some people who make money, who are dependent on the grid, who are going to be advocating for this protection. So it's nice for us to get together. We need the industry deep pockets. So we decided we'd reach out to them and go to the big data center industry conference, which was held in Orlando this fall. So I'm going to be reporting back on some of what we found out by doing that.

I also should say, we didn't just go—we were actually general session speakers. It was Dr. Baker, John Kappenman, and me. So we got to speak to the whole conference. People came up to us. We learned quite a bit. I'll also be briefly going through some of the official government reports. We've had a lot of experts talking about threats to the grid, but there is also a government consensus that I'll be briefly reviewing. Then Dr. Baker is going to go through a presentation about some specific threats to data centers. He actually has the incredible advantage of having consulted to the data center industry. So he brings a lot of knowledge.

And then we are going to have Mike Caruso from ETS-Lindgren. He also has real-world experience with the data center industry. So we hope to give you a view of what's happening in industry. And let's just think about data centers for a minute. I'll go through some of the critical sectors that are very dependent on data centers. We can start with the electric grid. We can move on to telecommunications. Telecommunications increasingly is dependent on data centers. These big switching centers are actually filled with computer servers. We had a presentation on healthcare earlier today with a captain from NORTHCOM. Well, medical records now are contained in data centers. Let's think about the Department of Defense—that's run on data centers, too. So you can see how the data center industry aspects that we are going to be talking about today really go across many of these critical infrastructures.

So these are some of the issues for data center operators. I'm going to tell you some complaints that they have with power quality issues as well as grid reliability. I'll give you a statistic: in the United States, the typical customer experiences one grid outage per year. So some of you may not have experienced one outage per year. And what that means is there are other people that are unfortunate that experience more than one outage. An amazing statistic—the average outage is about 200 minutes. That's a long time. Again, many of us don't experience outages of that duration but others that are less fortunate do.

It turns out that the United States actually ranks pretty far down the list in terms of electric grid reliability. We're behind many of the European countries in the reliability of our grid. It's interesting that when Congress passed the Energy Policy Act in 2005 they only mandated an adequate standard of electrical grid reliability; not a good or an excellent standard, just a an adequate standard. So these data centers have to deal with this unreliable electric power, and also poor quality. Also, they experience issues such as flicker, that's when the lights flicker; what the call voltage sag, harmonic distortion.

We heard presentations about how when there are these solar storms that can cause incredible harmonic distortion...the data centers have to deal with that. How do they do it? Well, it's expensive for them. They install very large, un-interruptible power supplies that have batteries. And then the batteries typically only last about 15 minutes for a data center. They are only designed to be a bridge until the diesel generators can kick in. Now, the problem with diesel generators, first of all, is that they are not reliable all the time.

So there is a range of reliability. One statistic that I've heard for nuclear power plants: their diesel generators are about 99 percent reliable. So they will start up about 99 percent of the time. And they have redundant generators. And so you do the math, the chance that two or three won't start up is quite small. But on the upper bound a statistic I've heard is, for example, a hospital or something like that, the generators wouldn't work about 30 percent of the time. So think about that.

Also, these diesel generators are hard to maintain because they are hard to test. You have to get environmental permits just to test the diesel generators. So you don't use the equipment every day, and it may not work when you need it. A big, big issue: there's only a few hours or days of onsite fuel for diesel generators. It's hard to store the fuel. It tends to age—you even get mold growing in the fuel or you get water condensing in the fuel. There is a process called "fuel polishing," where companies will come out and clean up your fuel for you. But I think it's appropriate to say that sometimes you may be dealing with contaminated fuel in a crisis.

Well, Hurricane Sandy showed some of the problems with these backup diesel generators for data centers. The data centers—there was a hospital in New York City, their diesel generators didn't go on. There was a data center, also in New York, where they actually had to have a bucket brigade with diesel fuel to resupply it. There were instances where the fuel trucks couldn't get through. So the data center industry is getting really, really frustrated with this situation. Many of them don't have back-up sites that are far afield that would be outside the outage area. A lot of times the back-up sites are within, say, an hour's drive—about 60 miles or so. That would be within the same control area.

The data center industry is now starting to clue in that you might have to be in a different control area or actually even in a different interconnection, as it's called. What is the data center industry doing? Well, they've tried to complain to their utilities for decades now and haven't made a lot of progress. Increasingly they're moving to their own generation. So data centers are now installing hydrogen fuel cells to go off the grid. These hydrogen fuel cells operate through natural gas—it turns out the natural gas system is more reliable than the electric grid, in many cases.

So that's a little bit of a summary of the situation with data centers. I'm now going to, just briefly, go through some of these official, U.S. government reports. Let's start out with the U.S. Department of Energy and the North American Electric Reliability Corporation (NERC). They came out with a report, "High Impact, Low Frequency Event Risks to the North American Bulk Power System." These high-impact, low frequency events, we've been talking about those all day. So even the people who supposedly protect the reliability of the grid recognize the threat. They've backed off a little bit on some of the conclusions of this report.

Then we have the National Academy of Sciences National Research Council. They produced a report in 2007—it was interesting, the Department of Homeland Security, when the report was finished, classified it. The authors of the report were very persistent, and in 2012 they succeeded in getting most of the report declassified. So now we can read it. We know the title: "Terrorism and the Electric Power Delivery System."

It's interesting. What it essentially says is if there was a physical attack on the bulk power system we could be without power days, weeks, months, likely even years. The U.S. Government Accountability Office has done some excellent work on the threat of cyberattack. The U.S. House of Representatives realizes that there is a problem with how electric grid reliability is regulated. I'll read a couple of the conclusions from this report because I think they are particularly pertinent. "Most utilities have not taken concrete steps to reduce the vulnerability of the grid to geomagnetic storms." And "Most utilities do not own spare transformers." Isn't that interesting? And if you look up a little bit, you can see "More than a dozen utilities reported daily, constant or frequent attempts at cyberattacks."

There is a report from the Department of Energy that talks about the difficulty of replacing transformers that had been damaged in a physical attack. It also talks about how these transformers, for the most part, are not manufactured here in the United States. The vast majority are manufactured overseas.

Here is a picture that shows the difficulty of replacing these large transformers. They can weigh up to 400 tons. They require special railway cars or special trucks. You see here that the overhead power lines sometimes have to be moved. It can take weeks to get one of these transformers transported to a location and replaced. And think about what would happen if an area was without power during that time.

Oak Ridge National Laboratory has produced a number of excellent reports on geomagnetic storms, on nuclear EMP. I should say all these reports are publicly available. They are on the Internet; anybody can read them. This one, authored by John Kappenman, is called "Geomagnetic Storms and Their Impacts on the US Power Grid." We have another report about radio frequency weapons, authored by William Radasky and Edward Savage. This is a lot of the analytic framework behind some of the EMP protective measures we've been hearing about today.

There is a really, big, significant study by the Congressional EMP Commission, talking about the threat of nuclear electromagnetic pulse, which could be a countrywide phenomenon. So look at all these government reports. And we've had experts here. I think that the experts we've had are credible. But these government reports are really, I would say, indisputable, in terms of a societal consensus that we have a very significant problem.

GEORGE BAKER: All right. I'm going to talk about big data center protection challenges, and the protection approach. And then I will get into some details there at the end. I was very disappointed Kevin Briggs couldn't be here because a lot of what we were talking about in the previous session, and this session—he's right in the middle of it. I saw his intended presentation. It was a good one, on the some of the challenges. And hopefully he will get permission to present it at the next go round.

So just, again, here is what I want to cover. An illustration. This is an Oliphant cartoon I thought was very instructive, where the lights go out and they realize the computers don't work. The computers are shut down. Somebody finds a flashlight and somebody needs to write a message and they don't know what "write" means because they are computer guys. And so, the bottom line is, we're doomed.

Well, this is what we want to avoid. And the good message is, you know, we're not SOL. There's lots of things we can do. We can avoid this, this kind of situation.

So the migration to Cloud computing has put more and more eggs in a smaller number of data center baskets. And people are migrating to larger and larger and larger data centers. And I've been to data centers where I'll say, "Well, where is your backup power?" And they'll say, "Well, we don't need it because if our site goes down, we've got another site, say, 200 miles away that will failover there. And we've got automatic failover. We've got logic circuit failovers." And I say, "Well, the EMP footprint is 1,000 miles. It's going to take out not just you but your failover site."

So this is a problem we need to watch very carefully. The EMP RF (radio frequency) weapon technology is diffused. I have a friend who got a Walmart vinyl briefcase and built one and burned out his own computer with it. He cut the antenna out of tin foil and used a fluorescent light starter for his power supply. The parts cost $200. That's just one example.

[One motivating factor is] the inadequacy of operational response procedures—you need to put in the physical protection. You're not going to have time to respond. I'm sorry. Just general lack of awareness is a problem, so one of the things we're trying to do is we give these presentations to the data center community and the electric power community, just to create awareness so people will know what embarrassing questions to ask their representatives and their power industry commissioners.

I'm not going to dwell on this but just basic EMP characteristics; we've already seen this in a previous talk: the power spectrum and the area coverage is the big problem. But notice that the EMP frequency spectrum only goes up to a gigahertz down here. And you have both E1 and E3. E3 is the very low frequency effect. With RF weapons, here is a picture of a couple of them, this is one that—the one on the right looks like the one that my cohort recently built. But the environment effects are comparable to EMP E1 except that you are going up to 10 gigahertz.

The thing is, with the RF weapons, the energy falls off this R^2, so the effective range is much more limited than with EMP, where you are getting thousands of miles. There are lot of people who, they think, yeah, you need to protect against RF weapons. But they don't see the RF weapons as an existential threat. You're not going to get the entire continent simultaneously like you would with a nuclear event. And so if you look at this, here is the frequency band. You can see the RF weapons are up there in the gigahertz range. Actually, we found that they are very effective down to 10 megahertz.

And here is a notional chart of the range to effect versus weight—the lower curve is damage, the upper curve is upset. And this is based upon a scalable ultra-wideband device that we developed. It's not all-inclusive but it just gives you some notion of how much weight it takes. If you want kilometers of range, you are talking about mounting things on trucks. Briefcases will have effective ranges of around a meter.

This is just—I tried to put on one chart the combined effects here. You've got the late, the EMP E3 and the geomagnetic disturbance in the first column there. The fast EMP E1 on the second column, then RF vulnerability. You can compare and contrast their system effects so that the grid power transmission distribution, the EMP is the most effective. The RF weapons can take out the control systems—that is a worry. But I don't think the RF weapons will take out the big transformers.

And you just look down the list there. The asterisk means that those systems are vulnerable to harmonics. There are certain things that will happen in data centers due to E3 because of the harmonic distortion on the power grid. But there you have it, just sort of compare and contrast all these different effects.

We know how to protect. I won't dwell on this because I talked about this in a previous presentation, but we know how to protect data centers. We've been doing it. And I should also mention, this is the protection approach that will be used for the FirstNet. And we know that the protection is affordable. We've been doing it. But you'd like to have the protection included in the system design, not as a retrofit.

And then the other, big, important thing that we stress when we talk to the data center owners, talking Microsoft and Google, Facebook, these companies have data centers—you need to really disperse your backup failover sites if possible, you know, on different continents. In many cases they are, they actually have their failover sites over on the Pacific Rim. That's good. But if you have all your failover sites in the continental United States, you could be in a world of hurt. I'll stop there.

MIKE CARUSO: Okay. I'm with ETS-Lindgren. We are a company that deals primarily in the passive control of RF energy. We've been around for about 50 years. I've been in the business about 31 years. And as Dr. Baker had talked about earlier, for DoD, for the past 30 years I've been involved in protecting military and government sites from HEMP (high altitude nuclear EMP). What we've noticed lately, over the past two years, is a very high increase in data centers looking to do EMP protection. And we found that to be very interesting, so I'll talk a little bit about it.

The data centers are opting for the global protection, as Dr. Baker was talking about in his last presentation. Talking about data centers, this is something that is very interesting. These are standards that are put out by the Uptime Institute—people that are heavily involved in data center evaluation and maintenance. In a Tier III data center, which, more likely than not, has some backup power and some backup capabilities, the allowable downtime per year is 1.6 hours. So that falls into the realm of the 200 minutes that you were talking about.

However, there are Tier IV data centers and those Tier IV data centers primarily service the financial industry and insurance and some of the very critical infrastructure elements that exist. Downtime for that is 2.4 minutes per year. That's the acceptable downtime. So it's something important to remember that for the data centers we're looking at protecting not only the power grid, we're protecting the data center itself. The equipment within the data center is quite vulnerable.

This is yet another example of a truck-mounted device. It can produce tens of thousands of volts per meter. Put the cover on the truck and drive it around and turn off the lights. So the data centers are not only aware of the EMP threat but they are keenly aware of the IEMI threat, the intentional electromagnetic interference threat. And the IEMI is the term that the IEEE (Institute of Electrical and Electronics Engineers) has adopted to talk about the high-power electromagnetic interference that would be created by a weapon—an RF weapon-type device.

This is kind of a sobering slide because it tells you the vulnerability of the standard equipment. Standard IT equipment, according to global harmonized standards for electromagnetic compatibility only gets tested to 10 volts per meter. After that it's vulnerable. And we're talking about thousands and tens of thousands of volts per meter in an EMP event or an IEMI attack from a truck-borne device. But even a suitcase-mounted device can exceed the 10 volts per meter. So if someone were to enter a data center or get close proximity to your equipment, at 10 volts per meter, it could interrupt it, probably not destroy it, with a suitcase device.

But going down the list, medical equipment we talked about before: 10 volts per meter. Telephone network equipment: 10 volts per meter. Aircraft cockpit controls: 7,200 volts per meter. We far exceed that with EMP. Automobiles, they are a little bit more stringent. They are 100 volts per meter. That's tops on them. And for military equipment we go to MIL-STD-461, 464: 200 volts per meter is the outside testing that happens on military equipment if it is not EMP-hardened.

This is just a depiction of a data center being protected. And looking at a data center—when I'm talking to data center people I'm talking to them about two modes of protection. With this, I'm talking about survival mode. In this mode we just go ahead and put a shielded enclosure in place. All the points of entry would be protected. The communications would be fiber optics. All of the precautions would be taken. But there would be no backup systems. So this, in a co-location data center, it would be sitting there. If the power went down, if the cooling went down, this data center would go down in an event. However, the equipment within the data center would stay protected and be able to be restarted when the services were brought back.

The second mode of protection is actually providing a secondary shelter, that houses backup power and backup cooling systems and cooling plants, that gets interconnected with the data center. This is for survival and continuous operation. Many of the financial institutions that I'm talking to on a regular basis are opting for continuous operation because offline they can lose millions of dollars per minute.

And as Tom was talking about, you take this to an economic level, you take this to basically buying insurance for their continuing operations, their disaster planning, their disaster recovery. And this is where it starts to make a lot of sense for these operations. For protecting an entire building, the added cost for the building alone, not including the equipment that might go in that building, but you are looking at about a 15 percent premium to go in and put the protection in and design it in up front as a new installation. If you're looking at a retrofit, then that cost is going to rise, somewhere between 20 and 30 percent as a retrofit because you've got to go in and do a lot of moving around.

Remembering that on data centers, the key is power and cooling. When protecting data centers, we've got big old water chillers sitting out on the roof. We've got generators outside. We've got racks of communications equipment. All of that needs to be protected. So, again, it's not just the power grid. So we've been talking about protecting power grid here, but we're talking about protecting critical infrastructure users with the data centers. It's becoming more and more important and a significant part of looking at the overall protection of the critical infrastructure.

MODERATOR (CHUCK MANTO): Very good. Thank you very much. It's really important, as you are thinking about microgrids and how do you protect them and the controls and everything you have for them, it is very much like protecting data centers. It is almost like having a little data center.

MIKE CARUSO: I'd like to clarify something. The 2 to 5 percent I quoted was for the total system cost. The 15 percent was just the building. It turns out the best data I've seen is, once you build a building, the cost of the electronics in the data center is eight times the building cost.

Maine Legislation on Power Grid Protection from Manmade and Natural EMP

http://youtu.be/AwQo-L1cZNk

Maine State Representative Andrea Boland

MODERATOR (CHUCK MANTO): When initiating discussions of disaster planning, if you don't want to shut someone down emotionally within the first few seconds of talking about something overwhelming, you give them a sense of hope. Because if you don't, you are going to lose them. One of the words I've often used is the word "triage." You know, there is one word that says there is something you can do really fast that doesn't cost much. So it's not as bad as you would otherwise think. There are some other things that we are just going to forget about because it is just too expensive. A lot of stuff in the middle that we can chunk up in a piecemeal fashion and prioritize.

So "triage" is one of those words for hope so that people don't emotionally shut down prematurely. So another word for hope right now is "Andrea Boland." Thank you, Andrea.

ANDREA BOLAND: What an introduction, huh? I've asked Tom Popik to stay up here with me because he was in attendance for three out of four of the hearings that we had in Maine on this, on my legislation. And he's made an effort to give me some notes so I don't wander too far afield. It's all been so interesting, I must say. Essentially, I get credit for passing the first legislation in the United States on EMP (electromagnetic pulse).

As you can imagine from the quality of the speakers you've heard today, I didn't do it singlehandedly. But I credit myself for the good sense to follow advice as I went along. A friend of mine, a scientist who I've relied on a lot for advice, put me onto this issue. And frankly I had enough things on my plate already. I wasn't excited to hear about this. I was horrified. But on the other hand, somebody had to do it, right? And when I heard that there was a problem in Washington, that was surprising. Why not Maine? We can give it a try.

So I went forward with that and it was really quite a lot, to take it on. But I was impressed with the urgency of the problem, in the first instance by Dr. Kirk, who helped open my eyes to this, and then by everything else that I read. I'm going to have a hard time following your notes, Tom.

Basically, what I did when I heard about this, and this is just to give you some information. Essentially what I wanted to do, because this was such a new issue to me, and I knew it was going to be new to the whole legislature, basically, was to learn as much as I can. And the first advice I had from Dr. Kirk was to talk to Dr. Pry. And, of course, I don't know any of these names.

I found Peter Pry and he was wonderful and forthcoming. It was exciting to talk to a man with a lot of vision and years of having worked really hard on this. And he told me then to talk to Congressman Roscoe Bartlett's office. And Bartlett's office opened up a whole, extra layer of information. Well, the thing that really impressed me was how enthusiastic and forthcoming these experts were—with their incredible experience and expertise—how willing they were to open up and share with me their gifts that they had to offer; one right after another.

In the first instance, and some of them are here in this room, obviously, Tom Popik, Cynthia Ayers, Bron Cikotas, Dr. Kirk, Peter Pry was here earlier, Bill Harris, and Dr. Baker. Anyway, so many wonderful people, and more. The main thing was to put this thing over.

So first, for me to be comfortable taking this on, I needed to learn more. So I proceeded to try to learn more from these people. They were very generous, they sent me papers and letters to share with the legislature. Because of the urgency of it, I wanted to bring it up as soon as I could. So I tried to bring it up in 2012 and that was a very short session. The leadership said, "It's too big. We can't take it on in this short session." But I shared a lot of information with them.

Peter told me I needed to come to Washington and convene a meeting with my delegation and others from the Department of Homeland Security. We did that. We met with Chris Beck from the EIS (Electric Infrastructure Security) Council. And it was considered a successful meeting. After that, I was really happy to be invited to the London EIS summit. I went on to that, which opened—just layers and layers were opening. I came back from there much more confident that I could bring this bill successfully—or at least be successful in informing the legislature.

What we did then was address the whole committee process. The following year, the leadership could not block any legislation. Whether they liked it or not, it had to go in to the process, and it would be heard by a committee. So we had the public hearing, we had some of these great folks come to testify. I got congratulated for that but really, the vision and the enthusiasm and the willingness of the experts to come forward and come to Maine of all places—I mean that's pretty far from Washington, D.C., in many ways. And Tom had to be persuaded a little harder than some, I guess, and he was just in New Hampshire.

We have to understand that these people have a lot of things to do besides chase around the country. So if you're looking in your legislatures to get something done, I think it is really important, one, to not ask someone in leadership, necessarily to bring it, because it's hard for leadership to take on a big, powerful industry in a new way. It's better to get somebody who is willing to work hard for it and just do the work and go along and let them sort of be brought along without having to be a great big target of industry.

So we had a fantastic public hearing. Bron Cikotas was there, too. Mike Maloof was present. The Energy, Utilities and Technology Committee—I was prepared to really have this thing kind of blown off. And I think that impression was given that, from the utilities, they are just going to treat it as, "Eh, another one of those Boland things," because I've gone against some other tough industries, too.

They tried to ignore it at first, so they didn't send anyone of consequence to the public hearing, which was terrific because, really, pretty much any testifying of any importance was done by the experts. I just had to introduce the bill and then make my remarks. My remarks had to do with the vulnerability of the grid—how easy it is to take it down, how unpredictable it is and the fact that Maine was in kind of a particularly vulnerable place in the United States.

I also spoke to them about our responsibility to our citizens and not to worry too much about the arguments for industry. I also spoke about the economic benefits of having a state where there are protections, where businesses need to have reliable electrical power, and how that could be a real draw—everybody is looking for economic development. So those are the pieces. And it was low cost. The great thing about this, as horrible as the scenario appeared to be, was there were some answers, which I had discovered from meeting these people, that could be applied. And they existed right now.

Some were just sort of design fixes that could go a long way. Others were the equipment. But we were there to bring hope as well as the vision of horror. There was also a vision of hope that was within reach. But anyway, at the end of sitting for four and a half hours straight listening to these experts one after another, the committee was totally bowled over. They were just transfixed. I mean, you don't see legislators sitting there with rapt attention for that extended period of time. And they did it.

The interesting thing after that was that after the public hearing, the whole committee was saying, "Oh, we've got to do something." And the next thing was, "What do we do?" But we go from that to what we call a work session. And usually there is just one work session. In the case of this piece of legislation there were three work sessions because there was so much to it.

One of the great things that I credit Tom Popik with is that what he brought to the public hearing was detailed information about our state, the state of Maine, and also about New England. He brought it right, smack down to ground level, home base, this is what's going on here. And it wasn't a very pretty picture, really. But he had been able to do some really great research and bring out data. It wasn't like, "Good grief, we've got to do something," it was, "Here's what you've got to work with."

After that, he and John Kappenman together, but particularly Tom Popik, put together a list of questions to recommend to the committee that they pose to the industry. And they were questions that went right to the heart of the matter. What do you do in this situation? What can you do here? What's your modeling? What is your data on this and that and the other thing? All things that had not been shared with the committee or anybody else in the public before.

And they were kind enough to supply the answers and share it with the industry. They had established a lot of credibility, so the committee said, "Okay. We'll ask those questions." And they sent the out to the industry. The industry came back to the next session with their so-called answers. Maybe about a half to two-thirds of the questions had answers. Many of them were so feeble it was embarrassing to sit and hear. The committee just sat—they kept waiting for the utilities who had now discovered that they did have to send some more senior people than their lobbyists out to this fight. And they were there.

But they were looking for a robust response from the utilities. And, of course, they didn't get it. And as that work session went on and the questions were brought up and the answers were given, it seemed like their eyes just got wider and wider and wider. They were practically throwing up their hands by the end. They just couldn't believe how unprepared Maine was, and how little they had for a plan to protect the grid and to respond to an emergency.

They started out being totally against this bill. And the position from ISO New England and our utilities, Central Maine Power, Bangor Hydro, was, "Don't worry, we've got this all under control. We've been doing fine for years, we will continue to. We've got operational strategies that will be fine. They've always served us just great." But, of course, when they had to answer these questions, that sort of reduced the potency of that argument.

And then they got to a position of, "Well, we're neither for nor against." You can be "for" or "against" or "neither for nor against." So they started out "against," and moved to "neither for nor against." But I credit the committee for giving the experts all the time they needed. The chair of the committee, and I credit myself for this because I had given them so much information, he knew there was something coming. That he didn't just say, "We're going to limit the testimony to 10 minutes or 15 minutes." He said, "I'm going to let your people have all the time that they need," which was wonderful, and they took it and did great things with it.

The other part of it that I credit Tom Popik and John Kappenman a lot for is they went and visited with the PUC (Public Utilities Commission). They talked to the people from the utilities companies and they worked up kind of a nice, collegial kind of environment to talk with them about these issues. And I think that made it a little easier for them to swallow some of the things that were being exposed about what they had to do, and what they had for answers.

In the end, the bill passed the committee unanimously but, of course, amended as we always do in the legislature. We don't leave anything alone. But it was okay for it to be amended because it was amended in a form that told the PUC, "Get all this information together. Look at the vulnerabilities for Maine. Look at the options for protection. Look at the cost: low-, middle- and high-cost strategies. Who's going to pay them? How is it going to work, policy-wise? Watch what's going on at FERC (Federal Energy Regulatory Commission) and NERC (North American Electric Reliability Corporation) throughout this and come back with a report in January and give us an opportunity then to report our permanent legislation with a plan."

So that's where we are right now. A lot of information has been given to them, and since then I know more information is being developed. What am I missing, Tom?

TOM POPIK: Tell them the final vote.

ANDREA BOLAND: Well, [before I tell you the final vote,] first I have to back up one place to credit myself a little bit on this. In order for you to be mindful of that when talking with legislators that may be thinking about this in your own states, it really is important because the industry has a position of thinking, they don't want to do anything about this, really. And they've been resisting it for years, many years. And they say, "We're working on it. We're working at something at the federal level. We're getting to it. Just hang tight, everything is going to be fine."

There's that problem to deal with, then there is the problem of these committees of jurisdiction being used to thinking of them as the authorities, the people who know something. And now, even though it had been exposed how little they really did know and how little prepared they really were, there is still that thing built into the heads of probably most people: "Well, you turn to the utilities for the answers." And so that is something that I had to work at getting past. And so, I really kept up with the members of the committee between these sessions and talking about it and reminding them of what they heard.

The other thing was, there's an analyst who had to write up what this law was going to be that they were passing, this bill. And she's an analyst in a non-partisan department of the legislature who has the job of just doing this sort of thing. So you do it legally. You make sure everything is according to Hoyle. But she sets it up according to what the chair's said. You have to birddog that, too, because she will be getting visited by the utilities, by the PUC, by others who have an interest in skewing the process.

And the part that Tom likes to say is that in the full legislature, in the Senate, the vote was 32 to 3 in favor of the bill. And in the House it was unanimous. So that was terrific.

But I just wanted to just say, it's just important to birddog it all the way because the industry is fighting you all the way. And even though at the end of these hearings and work sessions, between the work that we did and Tom and John did because they were there, and the chair of the committee, Representative Barry Hobbins did, in working to make everybody sort of play nice together and feel nice and collegial—I think we got more out of them, out of the utilities than was anticipated that we would.

And finally, in the last act of this, the chair of the committee said, "Now are we all in agreement that this is a good thing to do?" He polled all the people from the utilities as well as the experts in attendance and, of course, me. And everybody very nicely said, "Yes. This is a good thing to do." They are probably fighting it now, but you have to just sort of keep after it. We've gotten this far and I'm very excited about it. And I just want to thank Chuck for inviting me to the Dupont Summit last year and again this year, and the EIS Council for inviting me to London and then to Washington.

When I went to London everybody was sort of wondering what a state Senate, as state Representative was doing there because they were big, international figures. And then the next year I was invited to be a speaker. So that was kind of a big jump. It was kind of fun. But anyway, that's where we are. I'm really pleased to have been able to be an instrument of sort of helping kick open the door for these experts to move farther.

And then, having done that we could bring it all down to the National Conference of State legislatures in Atlanta with the help of Frank Gaffney's group, the Center for Security Policy in August and present it live, in person, to state legislators because as wonderful as all these experts were, none of our media wanted to report about it in Maine. You just couldn't get them to move. These are friendly people. I think the reporters would have liked to. One was moving on it, he thought, and then he got stopped. People aren't getting the information. It's not out in the public. People don't know about it.

So having done something in just as state, I could then bring it down to the NCSL (National Conference of State Legislatures), "How about if we do a panel? Can we do that?" We were able to do a panel. In the end it was pretty successful because we were able to get directly to state legislators who responded, and now we have a number of other states in the country who are moving forward with doing this. Shame on Congress that they had to wait for us to start pushing them, but if this is such an urgent issue, we've got to get it done one way or another.

And I wanted to just ask Tom if he had any remarks. And then after that I just wanted to close with a brief, four-minute video.

TOM POPIK: Sure. I'm going to be very brief. I'm going to say that Andrea Boland is a very modest person. Because not only did she have incredible legislative impact, she—as a non-technical person—has had nationwide technical impact. For those of you that were in yesterday's technical sessions, fully three of those presentations concerned data that Andrea, through her legislative process, had broken loose from the utilities. And so in the Emprimus presentation, talking about thousands of amps in the neutral during solar storms; that came from data that Andrea was able to get from the utilities.

ANDREA BOLAND: Tom was able to get from the utilities.

TOM POPIK: Oh, well, it was her process. And then, when we got the presentation from ABB, a lot of that had to do with refuting the data that Andrea had been able to break lose—as well as John Kappenman's presentation had to do with that data. It shows the power of what can be done at the state level. It is not just the laws at the state level. It's the technical impact that more disclosure of data can have. And so Andrea really deserves congratulations on two levels.

ANDREA BOLAND: Just one little sentence before we close with this video that I think is great, that the EIS Council people put together from one of the summits. Really, it was just such a pleasure to work with these people. And they are prepared to help so many others, if you just will listen to what they have to say and try to follow their advice. They just mean to be so kind and helpful throughout the country. And I know Frank Gaffney is working hard to coordinate a lot of this activity. So just remember that.

And one joyful thing about the video that I'm about to show you: we opened our panel at the NCSL conference to legislators with it, and we came right after the industry had been up there strong on their panel saying how there is not a problem. "We're doing fine. We've got operational procedures." They expected that they would start by picking up the phone and calling someone. Well, guess what. The phone might not be there. That's what operational procedures were about. But they had just been up there trashing the whole idea of doing anything about this.

And then we were able to open with this video and then go on to do more explanation to the legislators. So I hope you enjoy seeing it and thank you very much for your attention.

MODERATOR (CHUCK MANTO): While it is loading I'm just going to mention a few things. On this book in the back that talked about our presentation, on page 74 you will see Andrea Boland's question from the audience in one of the transcripts. And since that year, a lot has happened.

[Video clip from Electric Infrastructure Security Council]

MODERATOR (CHUCK MANTO): Thank you. And now a round of applause again for Andrea and Tom and crew.

Does anybody have any questions for Andrea—for example, about the process or other questions?

AUDIENCE MEMBER: I'm Bill Harris. I have a question for Andrea about the prospects for state legislation because the FERC process requires that federal standards go through NERC so they basically can block it. I know you had an adverse vote like a day or so ago. What are the prospects for June 2014? What do you think the states can do at the state level or cooperatively through this national conference?

ANDREA BOLAND: Well, what Bill Harris is referring to is this conference that I was at a couple of days ago asking for the National Conference of State Legislatures to support a resolution asking the federal government to support states that take initiatives like this and want to work on it. And at least to not get in their way if they do—and we haven't had anybody get in our way anyhow. There is an office within FERC whose mission is dedicated to supporting state critical infrastructure.

So, I had a resounding defeat, I have to tell you, on two resolutions that I brought. And I think it's too bad but it's hard for people to take in. There is a lot of information there. It is a little bit shocking. They weren't ready to put it into a policy. On the other hand, after they turned down those two resolutions, there was a third thing, which was an amendment to the policy, to the energy policy that had basically the same elements to it and they all voted for it.

So maybe there's hope that as information continues to be dumped on them they start to grow. But the good thing about this opportunity is that it was an opportunity to share information. And the more information we share, the farther we get. But essentially, the NCSL only as sort of a moral effect on the Congress to go and say, "All these states think you should do such and such." They may or may not go along with it.

MODERATOR (CHUCK MANTO): Yes, any other questions. Hold on for one second. Identify yourself please.

AUDIENCE MEMBER: Dr. Martin Dudziak, Virginia. And my question is simply, how can resources, including anybody in this room, join with people in the Commonwealth of Virginia who are trying to effect both legislative and civilian change for this? Here I am.

MODERATOR (CHUCK MANTO): Okay. Do we have somebody left from Virginia? But shy of Virginia, you may have a thought from Maine as to how Virginia might think of it.

ANDREA BOLAND: Well, I think there are a number of people in the room that are participating in outreach to other states. And as I mentioned before, Frank Gaffney of the Center for Energy Security was here. And he just left. He's one person. But anybody can contact me, that's for sure. And probably others whose cards you have picked up would be helpful in helping you network. But I, personally, don't have any contacts at this time.

MODERATOR (CHUCK MANTO): We have someone from the state of Virginia here who will identify himself and make a comment.

AUDIENCE MEMBER: Hi. It's Hank Cooper. And I live in Virginia and also in South Carolina. And Bob Newman, who spoke to us earlier, of course, lives in Virginia and he is a former adjutant general. I believe that the effort, which we are jointly beginning to pursue with the National Guard will have spillover contacts, which will get to state legislatures in both Virginia, South Carolina, and in between states as well.

MODERATOR (CHUCK MANTO): Let's have a comment from Mary who is from Maryland.

MARY LASKY: Yes. The national capital area is really important because that's where the government is. And if we do it in Virginia, that's great. If we do it in North Carolina that's great, or South Carolina. But it would be nice to do it also in Maryland so that we've got the national capitol area kind of covered with this. So if we could do it as a joint kind of adventure to go forth with this, I think that it would be really valuable. Any help that you can give us, Andrea, that would be great.

ANDREA BOLAND: Well, anybody can feel free to phone me but there are a lot of other people who can be helpful, as Ambassador Cooper pointed out, and some of the folks you have heard here. But anybody can feel free to call me. And I know you can catch these other folks, too.

MODERATOR (CHUCK MANTO): One of the things I wanted to mention as we do this is that there are different groups who are in this room. Some groups are organized to encourage specific actions because they are either lobbying organizations and they are allowed to do that; others are trade associations; others are standards organizations. InfraGard happens to be a very neutral, information-sharing organization.

And one of the things we are doing here is welcoming all of you to join InfraGard because it's free. You don't join as a company or an association. You join as an individual, and it's free. You just get a background check from the FBI, so it takes a little while, but then you are free to join these working groups that we are organizing nationwide in every one of these technologies and industries and policy arenas. And many of the leaders that you saw here today are involved in those.

And so when you talk to many of them, one moment they might have an InfraGard hat on as an information-sharing collaboration of these individuals. Another moment they may have a business hat on, and another moment they may have a legislator's hat on. And they are all different hats. We usually have more than one hat in the closet, right? So you are very much welcome to participate. And I like these questions because one of the things I like to do in these last few minutes is continue more dialogue and interaction from those of you here about ways you might want to do things in Virginia or wherever you are across the country.

And, by the way, we are still being webcast. So this is a public discussion and we have people from around the country who are listening to us now or watching us. And as it gets posted on YouTube and other places, you will be able to take what you saw today and share it with your colleagues and friends, and enlist their participation because we need information and ideas from them. So it is a true back and forth, collaborative information effort.

And I see another hand here. Identify yourself.

AUDIENCE MEMBER: Terry Hill with the Passive House Institute. Now, I don't know who could answer this question, but if you had a clean slate, given what you know now about all the dangers that we are facing, how would you design the grid now, with a clean slate?

MODERATOR (CHUCK MANTO): He's looking at me so I guess I get to answer that first, he's foolish enough to look at me when he said that. Actually, we bring a lot of people to the table who answer that question. And when we bring people from around the world, very often we look at people like from Europe who might say, "Don't build huge, regional grids. Build municipal grids." When you do things like create power, whether it's coal or anything else, and you have waste heat, capture that waste heat and sell the steam. So that the energy that you create locally will make you more money and you can do it more cost-effectively at a local level and be that much more resilient.

We all like the idea of centralized systems. And we heard today about microgrids. So the idea is to get as many people as we can from as many points of view all getting empowered to do power. That would be my way is to bring a lot more folks in. And Andrea, maybe you want to say something about that.

ANDREA BOLAND: Well, I just wanted to share what I've heard from others. I'm looking at Bron Cikotas—I know he always says this, too. That you need to have a plan and you need to identify what the most critical assets are that you want to protect, whether it would be, say, a hospital or a government building or a police station, a school. You might not be able to do everything.

And the other thing that was emphasized also is to protect these great big transformers, because they are so hard to replace. It takes up to two years to replace them—in good times. And if we've had a power outage, everybody is going to be in the market for them. So the line is going to be longer, even if they can get there. So those are the big ones that I understand.

MODERATOR (CHUCK MANTO): George has a hand up. So I'll give it to him before I make another comment and switch the questioning around.

GEORGE BAKER: The other thing I would add is, we would need to protect the data, the control centers. And we need to install more sensors so we're not flying blind. Right now we don't have the sensors that we need to detect the state of the grid in the presence of these threats. That would be important.

MODERATOR (CHUCK MANTO): As many of you know, we've done an economic impact assessment bounding the size of this problem. One of the outcomes of that showed that at least in the mid-impact scenarios, by protecting roughly 10 percent of the most critical infrastructure, you can avoid up to 60 percent of the economic losses. And so by selectively prioritizing like Bron said and Dr. Baker said, you can go a long way with relatively little. This goes back to this idea of triage. We can't do it all at once.

Concluding Remarks and Next Steps
for EMP SIG Working Group

http://youtu.be/arG-AWjvuG4

Mr. Chuck Manto, InfraGard National EMP SIG Chairman
Mr. Arnold Kishi, President of Hawaii InfraGard Members Alliance
Mr. Bron Cikotas, former executive from the Defense Nuclear Agency

CHUCK MANTO: I wanted to also bring a couple other people up to say at least a hello before we go. As we talked about this, I want to address the importance of what Andrea did. And, by the way, I think we have someone from Hawaii. I heard there might be an Assistant Secretary of Energy or past Assistant Secretary of Energy wanted to say hi. I don't know if they're all still here.

But notice what Andrea did by working systematically, persistently at all the right times, step by step by step, so it wasn't derailed. We need to do that across many of the states in the same way. What happened at NCSL (National Conference of State Legislatures) was not really a loss, because we had to actually continue to do that process, step by step, in the same way, in many other places. And I think we have an opportunity this coming year to do that.

This opportunity is not only in legislators, but with the users—users are stakeholders too, they are consumers and organizations, like hospitals. Imagine what would happen if we got a movement of hospitals who said, "We want to make certain hospitals can stay open indefinitely during a long-term power outage." Wouldn't that be huge? And let's say, if we did the same thing with firefighters, we brought out all the emergency management folks and the firefighters to say, "When there's a major emergency, we're going to keep the fire hall open." And, by the way, we know you don't have any money. Hospitals don't have much money. It's our job to find you the resources to do it.

If I could find you the resources to make your hospital more resilient, would you like that? If I could do that for your fire hall, would you like that? If I could do that for your 911 center, would you like that? Or would you turn down the money and say, "I'm too busy to talk to you right now. I don't want your money. I don't want to be able to make some of my own power or store it." That wouldn't make sense, right?

But we need to be able to engage them in a very positive way, saying, "We love you. We want to help," without saying, "I don't know that much about you. I don't really like you all that much. I'm just going to give you an unfunded mandate. And you've got to go suffer through this pain all by yourself."

Now unfortunately, terrorists give us a lot of unfunded mandates, as does Mother Nature. But I think if we worked together selectively, we can actually bring the resources to the table in a way that energizes

140

and motivates all these people to help us. Now I see someone, an InfraGard member. I want you to tell us your name and what you know about InfraGard and where you came from.

ARNOLD KISHI: Thank you, Chuck. I'm Arnold Kishi. I am the President of the InfraGard in Hawaii. But when I look at my career, my second home is actually the National Capitol Region, when you count all the years and days I've spent up here on the Hill, working with the federal agencies, this is my second home. So I understand a lot of the issues that you folks have been talking about in Virginia and Maryland very well.

Chuck has asked me just to say a few words as to why I am interested in all of this. Hawaii is a pretty isolated part of the country—and just some statistics, Hawaii is the 40th largest state, so there are 10 states that have less than a million and a half people. And, when you add all the land mass together, Hawaii is the 43rd largest state. So there are seven states that have less land than all the islands in Hawaii combined.

But what capitalizes the issues that Hawaii faces is that the geographers say it's the most isolated populated area in the world, because there's 2,500 miles to the next place that can provide assistance if Hawaii ever needs that. So it's, as a state and a community, developed a sense of being very self-reliant and does not take critical infrastructure for granted.

And, just as a personal note, when I was growing up, we had no electricity, no utilities, water was all catchment, and there was no sewer system. And telephone, when it finally came, was party line and only worked certain hours of the day. Same with the electricity. And in some places, that's still how it is. So I can appreciate a lot of the things that we do to keep the infrastructure running, but I always remember what it was like back when I was growing up. And, as I said, some places are still that way.

CHUCK MANTO: Thank you very much. Now, I wanted to tell you how important Hawaii is. You heard how significant and important Maine was. Let me tell you how important Hawaii is, from my point of view. In the event of one of these nationwide disasters, we are all Hawaii. We're all 2,500 miles away from the next person who could help you. That might be us, right?

So we need to be able to figure out how we can learn from Hawaii and be more like Hawaii in a lot of ways, and what we can do to be more like Maine. And I welcome all of your participation. Other questions? Let me take the one in the back and work my way forward. And identify yourself please.

AUDIENCE MEMBER: I'm Sandra Krebsbach and I'm actually here because I'm associated with technical education. But I'm one of these people who has another hat in the closet. I'm the mayor of a suburb of St. Paul/Minneapolis, and we have a transformer station. So this has become very interesting, and I just have a question in terms of how you manage discussion about this without residents who live near a transformer station becoming very concerned about their safety should there be an explosion.

CHUCK MANTO: Okay, I guess it depends on how close you live to the power station. But there's a similar question of how do you handle it when it goes bye-bye, and they're not getting the power? Who would like to take that? Our Minneapolis people, I think, had to catch a plane. So your neighbors had to go back. The Duluth fellow got snowed in—30 inches of snow, by the way—and I hear we're getting some freezing rain tomorrow. Anyone want to answer that? I got a couple. We'll do it in succession. Bron first and then Representative Boland after.

BRON CIKOTAS: This is Bron Cikotas. When we have seen those transformers fail, they usually do not fail in such a way where there's an explosion. So, from that standpoint, you're basically not in danger. But I have seen explosions of power substations—which are significant—on video. But again, typically, those things are not close to houses.

CHUCK MANTO: Do you ever hear that some of them are close to nuclear power plants?

BRON CIKOTAS: Yes, but the nuclear power plants are quite well protected from outside blasts. So I don't think that you have to worry about that. There are other things to worry about, but not that.

CHUCK MANTO: Thank you. It's a great problem to engage, and we don't have to be too fearful of that part. Andrea?

ANDREA BOLAND: I just wanted to say that those people are probably your greatest allies, because, you know, if they would engage, it's not just bad news, there's good news too. And you want to bring the good things into place.

CHUCK MANTO: Very good. I saw another hand here.

AUDIENCE MEMBER: Dr. Martin Dudziak, TetraDyn and ECOADUNA in Virginia. You mentioned about building alliances with different organizations. I have a thought, I just want to toss it out. Large-scale hospital corporations, HCA, Hospital Corporation of America, I know it quite well. Centera Corporation, Bon Secours. That's three right there. If we can craft a way to approach these organizations, particularly HCA, because I think we've got an ally—and I'm just tossing this out, I'm not speaking for retired former Senator Bill Frist, MD—but his family created HCA. When they went public the last time, a year and a half ago, it was the largest IPO in history, $33 billion. I think that we should get concrete and focus on the healthcare issues, because one of the biggest losses of life, after a large-scale EMP (electromagnetic pulse), will be as a consequence of food spoilage and contamination and other consequences related with food and water. If we can address getting an ally of such a corporation like HCA and others, then we will have power, economic, and political, I believe. Thank you.

CHUCK MANTO: Very good. Now when I hear you announced yourself as a doctor, is that a medical doctor or an academic doctor?

MARTIN DUDZIAK: For anybody who knows me, I'm too theoretical to work on anything clinical. I'm a theoretical physicist. And my PhD was partly on EMP.

CHUCK MANTO: Thank you very much. How appropriate. So my immediate reaction—I don't know if you've been here all day long, but one of the things we have done in recent days is we've begun to recruit top leaders in the country in various infrastructure areas. We have two co-chairs who are here today from the healthcare industry: one was Dr. Terbush who leads medical thinking nationwide at NORTHCOM (U.S. Northern Command), and the other is Dr. Terry Donat who is a thoracic surgeon from Chicago and is active in the InfraGard section there. Both of them have joined our EMP SIG (Electromagnetic Pulse Special Interest Group) and are co-chairing that. So we are absolutely interested in that.

And I would like to recruit your involvement personally to bring Senator Frist and the CEOs of those organizations to the table. And we will bring in the world's best experts in all these related areas to calmly and systematically provide them with the information they need so they could determine how they could be effectively and, if nothing else, entertainingly involved in what we need to do, to protect their own self-interest. So I'd very much like to do that.

And so what I'd like to do, because I have so many things in my head, is throw the ball back to you and now say, you now have hunting rights. You've been given a hunting license to hunt me down, and the others that you've met. And, if you can't find the others, find me. I'm very findable. I may be exhausted and tired and a little loopy, but you are very welcome to hunt me down and do that.

I want to mention food. Sometimes people get tired of hearing me say this. But Congressman Bartlett has not only agreed to be on our policy committee and our energy committee and our civilian military committee, he's very interested in working with us on a strategic proposal. That's a policy proposal that I'll just mention briefly, because you mentioned food.

During the Cold War, we had a Strategic Grain Reserve. At any point in time we could feed 50 million starving people anywhere in the world—it was basically to protect ourselves, right. And who did we normally feed? Who starved before us? Millions of Chinese, millions of Russians. And we used our grain to keep those people alive. Wasn't that wonderful?

But guess what: we don't have that anymore. So if 50 million Chinese or Russians today came to us and asked for aid, we couldn't feed them. What's worse, what happens if 50 million Americans are starving? We can print all the money we want but if there's no food, we can't feed ourselves.

One of the reasons why we don't have that is that it was a centralized system obtainable by the Feds. And so, whenever we had a bad crop year, and the poor farmers were saying, "Well, the prices are going up. Maybe I can make something out of this and not go bankrupt." To be a political hero someone in the federal government said, "Prices are going up. Wow. Bad. Inflation. Bad. I know—I have a tool to do something about it. I'm going to dump this grain on the market and suppress food prices."

And it did that, except for the fact that a whole bunch of farmers went out of business. And the grain industry is tottering. And so, they basically begged the federal government in public documents you can see, "Please, don't give us any more money. Please dismantle this system." And another hero, federally, said, "You're right. We have too big of a deficit. We don't want to spend money if we don't have to. We will just discontinue that program." So we no longer have it.

So I've been proposing, along with Congressman Bartlett and others, to tweak that a little bit. And, instead of having a centralized system that could be commandeered for political purposes and disrupt the market, what we instead do is have a distributed system that not only has grain, but any other kind of nutritional food. And we call it the Strategic Food Reserve. And we figure out a way to make it possible for us to feed all of America for virtually no extra money, and without a nickel from the government, in such a way that we can feed ourselves in case of an emergency.

We started thinking through, as economists and people in that industry, that there could be a very cost-effective way to do that, using basic market methods that could basically cost us almost no money. And we can talk about that later, but that's an example.

As we begin to work on the food issue, or the healthcare issue, we're all going to start off with a bunch of great ideas that, six months later, are going to look half-baked, right? But if you're an entrepreneur, or you're an inventor, you realize that almost every decent idea that made somebody a fortune or did something wonderful started out as a half-baked idea.

But, when you sit in a room, and you listen to your fellow technical people, and your marking people, and your manufacturing people, and your customers, and you get a heat of criticism from your peers fast enough to do you some good, that heat of that constructive criticism takes your half-baked loaf of bread and makes it a fully-baked loaf.

And that's why we need each of you. So you are now drafted by me—and if you don't show up, you're AWOL, and I'll send the army guys after you. Okay? Anybody else who dares to have a question? Do I see another hand somewhere? Oh Mary—Mary is one of the people who is one of our three support people we have to help you in this. One is Mary, because she is helping support all these liaison panels that we have.

A second person is Lauren Schuler from the FBI. She is our full-time FBI agent in the headquarters of FBI, to support our efforts. And then we have a third, we have a national InfraGard board member who is going to help outreach to all of the InfraGard organizations including Hawaii. And his name is Bob Janusaitis, to make certain that folks like him are brought to the table as we need to have that happen. Mary?

MARY LASKY: We still have, out at the front desk, the sign-up sheets for being part of these working groups. So if you are motivated from what you've heard today, and will come and help us, it would be wonderful if you'd sign up. And then the leaders of those various areas are going to contact you.

CHUCK MANTO: Okay, now, another thing I want to mention, if you believe in resiliency, that means you always have a plan B, right? Okay. So our InfraGard website is being redeveloped right now for us. And it's not as fully functional yet as we might like. I know that never happens with a federal IT project. However, that's even when times are going well and we got everything under control. But you never know.

So one of the things we're offering is our email addresses, on the back of that book I showed you that you get to take with you today. The story of what we're doing is all on the back flap—our mission, our goals, and a way to hunt me down is on the back flap of that.

Secondly, we will try to do something with things like Eventbrite. We have a communications team, and our communications technology people are going to explore ways to facilitate interaction, nation-wide, on a regular enough basis so you can feel empowered to do something. We have people who are joining us who are experts at social media. We have, for example, people here today who are developing technology that will make it possible for any of us who are not computer-savvy, to download a little piece of software on our computer, that would allow us to make our own mobile app on the fly—without knowing how to do any programming whatsoever—just by dragging and dropping things.

We have a mobile app today supporting this. If you go to Eventbrite and download the Eventbrite application (it's a free app) and all you have to do is search on their search bar, when it comes up on your phone or iPad. You input the letters EMP and guess what? This meeting pops up with all of the other links and everything else. We will create an InfraGard National EMP SIG app as well, so that we can continue to find ways to talk with each other.

We want to make it possible to do one-on-one communications like this, and within groups. We want to be able to make it possible for those of you in your groups locally, wherever you are in the country, or with the people in your subject matter areas so you can work amongst yourselves. You don't have to wait for this conference to happen, and you don't need to wait for me, for sure, or anyone else in the group. We're trying to facilitate that communication.

Anything else before we go? Because we're at the time to close. Oh, perfect, I hear a voice. Yes, Denisa. This will be the last comment or question.

DENISA SCOTT: So Chuck was just mentioning about the drag and drop interface. And I am actually one of the partners in the company that has created this framework. And one of the things that I would like to do is find people who are interested in creating something useful and productive, constructive, that would actually be relevant to this particular SIG, to the kinds of requirements that may emerge, both for preparation and for response purposes. So, if you are interested, please contact me.

CHUCK MANTO: Great. Before we leave, I wanted Bron to be able to make sort of a closing remark or two, to sort of pull everything together, because you never know when we may have missed something. Dr. Pry was going to do that for me, but he left, so I forgot about it, and on the fly, plan B, Bron agreed to do it. And so this is Bron's opportunity to make certain to give us a closing overview or remark or whatever.

BRON CIKOTAS: Thank you. This is a big problem, we all realize it. But the basic thing is, we do not need to harden all our infrastructures. In my opinion, if we hardened 30 percent of our grid, we could save a majority of our people. To do that, you basically need to prioritize. We need to harden the nuclear plants. We need to harden the hydro plants. They do not need fuel supplies for a long time. So that type of prioritization needs to be across the board.

We need to harden certain infrastructures like water, like sewage processing, plus others that are critical, like transportation and things like that, that are critical to both maintaining life at a level that people can survive, and at the same time, to initiate the recovery process.

So, if you look at the whole problem, you say there's too much to do, but there are ways of addressing it that, from the start, you should start saving lives, right from the beginning. You don't wait to get the plant finished and then start. It's going to take time to do it, but we can start, and we can start saving lives right from the beginning. And that is the key.

I'd like to leave you with one more thing—this is probably new to most of you. I've probably done more EMP testing than anybody else. I've been working in this field since 1963. In my experience, I think I'm the only one that's run a test like this. I set up a room half the size of this hall, set up military operators to operate equipment as we were testing it. And we were exposing those people to 50,000 volts per meter with the equipment operating.

Now I have been exposed to those levels many times before, so I didn't mind putting people in that condition. We made sure nobody had a pacemaker or anything like that. But basically, as we were pulsing out at 50,000 volts per meter, we had the operators operate keyboards. They were getting shocked at the keyboards. We had the test conductor wearing one of those headset radios. He got zapped from the ceiling. And that was a very unpleasant experience, but there was no serious injury in that case.

The reason I'm bringing this up to you, is that as we were pulsing, every outlet in that place was sparking. There were arcs in every outlet or arcs in other locations within that particular test setup. What that means is that in your home, if you get exposed to high-altitude EMP, you're going to have sparks occurring across the home. In our EMP testing, we started a fire at an AT&T facility. Now, it's not easy to start a fire in AT&T facility, but we did.

Basically, we have to be prepared to deal with those arcs. As I go around talking around the country on these issues, on hardening—I still work in that area—I basically recommend to people that what they do for their own protection is get these power strips, high value equipment power strips, and either plug them in throughout their house, or else there are units that are available where you can take a regular outlet, replace it, and it will provide the protection. The idea is not to get a fire started in your house. You put three or four or five of those in your house, basically you should not have any unintended sparking locations, arc-over locations.

The problem is, if you live close to your neighbors, you want them to do it also. Because if their house catches on fire, you might just get yours burned down too. So it's an issue that people generally haven't talked about. You rarely hear it. I think I'm the only one talking about it, and it's one that is serious. It could start serious fires, in cities particularly, when there are no traffic lights. Eventually, you run out of water. And it's a situation that we need to carefully look at. The nice thing is, if we could get the national electrical code to put those things throughout houses in the cities or in the countryside, we could avoid this problem.

CHUCK MANTO: Thank you very much. And I've been told that we're running out of contract time. We may be running up to union laws, I don't know. Thank you all for coming. We also are organizing an organization looking at all the standards bodies. But thanks again. Give yourselves a round of applause as you say goodbye.

AUDIENCE MEMBER: Chuck has done such an amazing job here. I don't know what zapped him to have all his cells sparking like that, but let's have a round of applause for Chuck.

2014 Virginia Legislative Session

Senate Joint Resolution No. 61

Directing the Joint Commission on Technology and Science to study strategies for preventing and mitigating potential damages caused by geomagnetic disturbances and electromagnetic pulses. Report.

Agreed to by the Senate, February 11, 2014
Agreed to by the House of Delegates, February 19, 2014

WHEREAS, geomagnetic disturbances and electromagnetic pulses have the capability of producing significant damage to the Commonwealth's infrastructure and electronic equipment; and

WHEREAS, the Commonwealth's vulnerability to such threats is increasing daily through heightened use of and dependence on electronic equipment; and

WHEREAS, the Joint Commission on Technology and Science may be able to identify measures to protect the Commonwealth's infrastructure through focused examination; now, therefore, be it

RESOLVED by the Senate, the House of Delegates concurring that the Joint Commission on Technology and Science be directed to study strategies for preventing and mitigating potential damages caused by geomagnetic disturbances and electromagnetic pulses.

In conducting its study, the Joint Commission on Technology and Science shall (i) study the nature and magnitude of potential threats to the Commonwealth caused by geomagnetic disturbances and electromagnetic pulses; (ii) examine the Commonwealth's vulnerabilities to the potential negative impacts of geomagnetic disturbances and electromagnetic pulses; (iii) identify strategies to prevent and mitigate the effects of geomagnetic disturbances and electromagnetic pulses on the Commonwealth's infrastructure; (iv) estimate the feasibility and costs of such preventative and mitigation measures; and (v) make recommendations regarding strategies that the Commonwealth should employ to better protect itself from and mitigate damages caused by geomagnetic disturbances and electromagnetic pulses.

All agencies of the Commonwealth shall provide assistance to the Joint Commission on Technology and Science for this study, upon request.

The Joint Commission on Technology and Science shall complete its meetings by November 30, 2014, and the Chairman shall submit to the Division of Legislative Automated Systems an executive summary of its findings and recommendations no later than the first day of the 2015 Regular Session of the General Assembly. The executive summary shall state whether the Joint Commission on Technology and Science intends to submit to the General Assembly and the Governor a report of its findings and recommendations for publication as a House or Senate document. The executive summary and report shall be submitted as provided in the procedures of the Division of Legislative Automated Systems for the processing of legislative documents and reports and shall be posted on the General Assembly's website.

Source: Virginia's Legislative Information System
VIRGINIA LAW PORTAL
http://lis.virginia.gov/cgi-bin/legp604.exe?141+ful+SJ61ER

126ᵀᴴ MAINE LEGISLATURE, FIRST REGULAR SESSION

Resolve, Directing the Public Utilities Commission To Examine Measures To Mitigate the Effects of Geomagnetic Disturbances and Electromagnetic Pulse on the State's Transmission System

Emergency preamble. Whereas, acts and resolves of the Legislature do not become effective until 90 days after adjournment unless enacted as emergencies; and

Whereas, the North American Electric Reliability Corporation has identified 2013 as a peak year of solar activity that could result in a geomagnetic disturbance; and

Whereas, the impact of a significant geomagnetic disturbance or electromagnetic pulse on the reliability of Maine's electric grid is unknown; and

Whereas, the Public Utilities Commission may be able to identify measures to protect Maine's electric grid through a focused examination; and

Whereas, in the judgment of the Legislature, these facts create an emergency within the meaning of the Constitution of Maine and require the following legislation as immediately necessary for the preservation of the public peace, health and safety; now, therefore, be it

Sec. 1 Examination of vulnerabilities and mitigation. Resolved: That the Public Utilities Commission shall examine the vulnerabilities of the State's transmission infrastructure to the potential negative impacts of a geomagnetic disturbance or electromagnetic pulse capable of disabling, disrupting or destroying a transmission and distribution system and identify potential mitigation measures. In its examination, the commission shall:

1. Identify the most vulnerable components of the State's transmission system;

2. Identify potential mitigation measures to decrease the negative impacts of a geomagnetic disturbance or electromagnetic pulse;

3. Estimate the costs of potential mitigation measures and develop options for low-cost, mid-cost, and high-cost measures;

4. Examine the positive and negative effects of adopting a policy to incorporate mitigation measures into the future construction of transmission lines and the positive and negative effects of retrofitting existing transmission lines;

5. Examine any potential effects of the State adopting a policy under subsection 4 on the regional transmission system;

6. Develop a time frame for the adoption of mitigation measures; and

7. Develop recommendations regarding the allocation of costs to mitigate the effects of geomagnetic disturbances or electromagnetic pulse on the State's transmission system and identify which costs, if any, should be the responsibility of shareholders or ratepayers; and be it further

Sec. 2 Monitor federal efforts regarding mitigation measures. Resolved: That the Public Utilities Commission shall actively monitor the efforts by the Federal Energy Regulatory Commission, the North American Electric Reliability Corporation, ISO New England and other regional and federal organizations to develop reliability standards related to geomagnetic disturbances and electromagnetic pulse; and be it further

Sec. 3 Report. Resolved: That the Public Utilities Commission shall report the results of its examination required pursuant to Section 1 and the progress of regional and national efforts to develop reliability standards under Section 2 to the Joint Standing Committee on Energy, Utilities and Technology by January 20, 2014. The Joint Standing Committee on Energy, Utilities and Technology may submit a bill to the Second Regular Session of the 126th Legislature based on the report.

Emergency clause. In view of the emergency cited in the preamble, this legislation takes effect when approved.

HR 3410 Critical Infrastructure Protection Act

H.R.3410—CRITICAL INFRASTRUCTURE PROTECTION ACT

To amend the Homeland Security Act of 2002 to secure critical infrastructure against electromagnetic pulses, and for other purposes.

HR 3410 IH
113th CONGRESS
First Session
H. R. 3410

To amend the Homeland Security Act of 2002 to secure critical infrastructure against electromagnetic pulses, and for other purposes.

IN THE HOUSE OF REPRESENTATIVES

OCTOBER 30, 2013

Mr. FRANKS of Arizona (for himself and Mr. SESSIONS) introduced the following bill; which was referred to the Committee on Homeland Security

A BILL

To amend the Homeland Security Act of 2002 to secure critical infrastructure against electromagnetic pulses, and for other purposes.

Be it enacted by the Senate and House of Representatives of the United States of America in Congress assembled,

SEC. 1. SHORT TITLE.

This Act may be cited as the 'Critical Infrastructure Protection Act' or 'CIPA'.

SEC. 2. EMP PLANNING, RESEARCH AND DEVELOPMENT, AND PROTECTION AND PREPAREDNESS.

(a) In General—The Homeland Security Act of 2002 (6 U.S.C. 121) is amended—

(1) in section 2 (6 U.S.C. 101), by inserting after paragraph (6) the following: '(6a) EMP—The term 'EMP' means—

'(A) an electromagnetic pulse caused by intentional means, including acts of terrorism; and (B) an electromagnetic pulse caused by solar storms or other naturally occurring phenomena.';

(2) in title V (6 U.S.C. 311 et seq.), by adding at the end the following:

'SEC. 526. NATIONAL PLANNING SCENARIOS AND EDUCATION.

'The Secretary, acting through the Assistant Secretary of the National Protection and Programs Directorate, shall—

'(1) include in national planning scenarios the threat of EMP events; and '(2) conduct a campaign to proactively educate owners and operators of critical infrastructure, emergency planners, and emergency responders at all levels of government of the threat of EMP events'; (3) in title III (6 U.S.C. 181 et seq.), by adding at the end of the following:

'SEC. 318. EMP RESEARCH AND DEVELOPMENT.

'(a) In General—In furtherance of domestic preparedness and response, the Secretary, acting through the Under Secretary for Science and Technology, and in consultation with other relevant agencies and departments of the Federal Government and relevant owners and operators of critical infrastructure, shall conduct research and development to mitigate the consequences of EMP events. '(b) Scope—The scope of the research and development under subsection (a) shall include the following:

'(1) An objective scientific analysis of the risks to critical infrastructures from a range of EMP events. '(2) Determination of the critical national security assets and vital civic utilities and infrastructures that are at risk from EMP events. '(3) An evaluation of emergency planning and response technologies that would address the findings and recommendations of experts, including those of the Commission to Assess the Threat to the United States from Electromagnetic Pulse Attack. '(4) An analysis of technology options that are available to improve the resiliency of critical infrastructure to EMP. '(5) The restoration and recovery capabilities of critical infrastructure under differing levels of damage and disruption from various EMP events.'; and (4) in section 201(d) (6 U.S.C. 121(d)), by adding at the end the following: '(26)(A) Prepare and submit to the Committee on Homeland Security of the House of Representatives and the Committee on Homeland Security and Governmental Affairs of the Senate—

'(i) a comprehensive plan to protect and prepare the critical infrastructure of the American homeland against EMP events, including from acts of terrorism; and '(ii) biennial updates of such plan.

'(B) The comprehensive plan shall—

'(i) be based on findings of the research and development conducted under section 318; '(ii) be developed in consultation with the relevant Federal sector-specific agencies (as defined under Homeland Security Presidential Directive-7) for critical infrastructures; '(iii) be developed in consultation with the relevant sector coordinating councils for critical infrastructures; and '(iv) include a classified annex'.

(b) Clerical Amendments—The table of contents in section 1(b) of such Act is amended—

(1) by adding at the end of the items relating to title V the following: 'Sec. 526. National planning scenarios and education.'; and (2) by adding at the end of the items relating to title III the following: 'Sec. 318. EMP research and development'.

(c) Deadline for Initial Plan—The Secretary of Homeland Security shall submit the comprehensive plan required under the amendment made by subsection (a)(4) by not later than one year after the date of the enactment of this Act. (d) Report—The Secretary shall submit a report to Congress by not later than 180 days after the date of the enactment of this Act describing the progress made in, and an estimated date by which the Department of Homeland Security will have completed—

(1) including EMP (as defined in the amendment made by subsection (a)(1)) threats in national planning scenarios; (2) research and development described in the amendment made by subsection (a)(3); (3) development of the comprehensive plan required under the amendment made by subsection (a)(4); and (4) beginning a campaign to proactively educate emergency planners and emergency responders at all levels of government regarding the threat of EMP events.

Appendix

The following documents provide additional background material for the topics covered in these proceedings. A number of these materials and accompanying video recordings of presentations can be found on the Policy Studies Organization website beginning at this web address: http://www.ipsonet.org/conferences/the-dupont-summit. The appendix entries of public documents below are hyperlinked so that readers can gain direct access to the documents. For example, many of the considerable number of documents submitted to the Maine PUC are itemized according to their list of comments and filings linked in this bibliography but the complete collection can be obtained after registration by going to their website at: http://www.maine.gov/mpuc/online/index.shtml and looking up case 2013-00415.

Overview

Baker, Daniel; *"New Twists in Earth's Radiation Belts,"* American Scientist # Vol. 102, 2014—Rings of high-energy particles encircling our planet change more than researchers realized. Those variations could amplify damage from solar storms. http://www.americanscientist.org/issues/feature/2014/5/new-twists-in-earths-radiation-belts

Dobbins, R.; C. J. Schrijver, C.J: Murtagh, W.;Petrinic, M.; *"Assessing the impact of space weather on the electric power grid based on insurance claims for industrial electrical equipment"* Space Weather Journal, Wiley; July 2014. Authors review ten years of insurance claims linking damage to regular space weather effects. http://onlinelibrary.wiley.com/enhanced/doi/10.1002/2014SW001066/

EMP Commission documents (www.empcommission.org). This includes the Executive Summary Report of 2004 and the Final Report of 2008.

EMP MIL SPEC 188.125 for EMP Protection
https://www.google.com/#q=mil+spec+188.125

Kemp, John; Reuters, Feb 18, 2014, *"U.S. orders power grid to prepare for solar storms:"* http://www.reuters.com/article/2014/02/18/electricity-solar-storms-idUSL6N0LN3HU20140218

Resilient_Societies_Press_Release_June_16_2014_Transmitted-2
http://www.resilientsocieties.org/pressreleases.html

FERC Rulings and Publications

Rule 779 on Geomagnetic Disturbances, See 143 FERC 61,147, United State of America, Federal Energy Regulatory Commission, 18 CFR Part 40, [Docket No. RM12-22-000; Order No. 779], Reliability Standards for Geomagnetic Disturbances (Issued May 16, 2013) http://www.ferc.gov/whats-new/comm-meet/2013/051613/E-5.pdf

FERC Remand of NERC Cyber Regulations
    http://www.ferc.gov/whats-new/comm-meet/2013/032113/E-11.pdf.
    142 FERC 61,204, United States of America, Federal Ener-
    gy Regulatory Commission, Docket No. RD12-5-000

FERC, Executive Summary, Effects of EMP on Electric Power Grid
    http://www.ferc.gov/industries/electric/indus-act/reliabil-
    ity/cybersecurity/ferc_executive_summary.pdf

FERC, LaFleur, Cheryl A.; Acting Chairman before Committee on Energy and Commerce, Sub-
    committee on Energy and Power, United States House of Representatives, Hearing on the
    Role of FERC in a Changing Energy Landscape, Dec. 5, 2013.
    http://ferc.gov/CalendarFiles/20131205094201-LaFleur-12-05-2013.pdf

"JOINT MEETING OF THE NUCLEAR REGULATORY COMMISSION AND THE FEDERAL
ENERGY REGULATORY COMMISSION" AD06-6-000, Friday, June 15, 2012, 9:30-11:30
a.m. (covers discussion between FERC, NRC and NERC on solar storm effects on the grid.
http://www.nrc.gov/reading-rm/doc-collections/commission/tr/2012/20120615.pdf

Nuclear Regulatory Commission

Long-Term Cooling and Unattended Water Makeup of Spent Fuel Pools, NRC-2011-0069 (Phased rule
    making in response to Petition from Tom Popik on GMD impacts on ability to cool spent fuel rods.)
    http://www.regulations.gov/#!documentDetail;D=NRC-2011-0069-0109

Station Blackout Mitigation Strategies [NRC-2011-0299]
    https://www.federalregister.gov/regulations/3150-AJ08/station-blackout-mitigation-strategies-
    nrc-2011-0299-
    Summary: The NRC published an Advance Notice of Proposed Rulemaking (ANPR) on
    March 20, 2012 (77 FR 16175), to seek public comments on potential changes to the
    Commission's regulations that address a condition known as station blackout (SBO).
    SBO involves the loss of all onsite and offsite alternating current (ac) power at a nu-
    clear power plant. A central objective of this rulemaking would be to make generi-
    cally applicable requirements previously imposed on licensees by EA-12-049 "Order
    Modifying Licenses with regard to Requirements for Mitigating Strategies for Beyond-
    Design-Basis External Events," while ensuring that the new requirements are properly
    integrated with the existing SBO requirements in 10 CFR 50.63. This regulatory ac-
    tion is one of the near-term actions based on lessons-learned stemming from the March
    2011, Fukushima Dai-ichi event in Japan. Includes references to GMD impacts.

Also see: https://www.federalregister.gov/articles/2012/03/20/2012-6665/station-blackout#h-15

Significant Press Article on High-impact Threats

NASA Website on July 2012 Solar Storm Near Miss, July 23, 2014
http://science.nasa.gov/science-news/science-at-nasa/2014/23jul_superstorm/

Washington Post Editorial Board on Solar Storms, August 9, 2014
http://www.washingtonpost.com/opinions/extreme-space-weather-threatens-to-leave-the-us-in-the-dark/2014/08/09/22782cd4-1c26-11e4-82f9-2cd6fa8da5c4_story.html

Wall Street Journal Articles

Ryan, Tracy, WSJ, May 14, 2014, "Here Comes the Sun Storm,"
http://online.wsj.com/news/articles/SB10001424052702303505504577404360076098508

Smith, Rebecca, WSJ, Feb. 5, 2014; "Assault on California Power Station Raises Alarm on Potential for Terrorism"—'April Sniper Attack Knocked Out Substation, Raises Concern for Country's Power Grid.' http://online.wsj.com/news/articles/SB10001424052702304851104579359141941621778

Smith, Rebecca, WSJ, Mar. 12, 2014 "U.S. Risks National Blackout From Small-Scale Attack—Federal Analysis Says Sabotage of Nine Key Substations Is Sufficient for Broad Outage." http://online.wsj.com/news/articles/SB10001424052702304020104579433670284061220

Smith, Rebecca, WSJ, Mar. 4, 2014 "Transformers Expose Limits in Securing Power Grid." http://online.wsj.com/news/articles/SB10001424052702304071004579409631825984744

Bibliography of Comments Filed with the Maine Public Utility Commission Docket 2013-00415 for LD 131 in 2013

The Sage Policy Group, economic impact report, *Initial Economic Assessment of Electromagnetic Pulse (EMP) Impact upon the Baltimore-Washington-Richmond Region,* September 10, 2007

Ambassador R. James Woolsey, letter to State of Maine, *Maine Should Protect Its Electric Grid From Electromagnetic Pulse (EMP)*

R. James Woolsey, testimony, *Testimony Before the House Committee on Energy and Commerce,* May 21, 2013

Electric Infrastructure Security Council, comments for MPUC and attachments, *Inquiry into Measures to Mitigate the Effects of Geomagnetic Disturbances and Electromagnetic Pulse On the Transmission System in Maine,* October 4, 2013

Advanced Fusion Systems LLC, statement of capabilities, *Advanced Fusion Systems LLC Hardware and Test Capabilities,* October 4, 2013

Foundation for Resilient Societies, comments to MPUC, *Comments of The Foundation for Resilient Societies in Response to 14 Questions Propounded by the Public Utilities Commission of the State of Maine Together with Appendices,* October 4, 2013

Foundation for Resilient Societies, Appendix 1 of 4 to comments to MPUC, *Response to NERC Request for Comments on Geomagnetic Disturbance Planning Application Guide,* August 9, 2013

Foundation for Resilient Societies, Appendix 2 of 4 to comments to MPUC, *Comments of The Foundation for Resilient Societies Before the Federal Energy Regulatory Commission, Reliability Standards for Geomagnetic Disturbances, Docket No. RM12-22-000,* December 24, 2012

Foundation for Resilient Societies, Appendix 3 of 4 to comments to MPUC, *Comments of The Foundation for Resilient Societies Before the Federal Energy Regulatory Commission, Reliability Standards for Geomagnetic Disturbances, Docket No. RM12-22-000,* April 1, 2013

Foundation for Resilient Societies, Appendix 4 of 4 to comments to MPUC, *Comments of The Foundation for Resilient Societies Before the Federal Energy Regulatory Commission, Reliability Standards for Geomagnetic Disturbances, Docket No. RM12-22-000,* May 14, 2013

Emprimus LLC, answers to questions arising from MPUC Interim Report, *LD 131, Resolve, Directing the Public Utilities Commission To Examine Measures To Mitigate the Effects of Geomagnetic Disturbances and Electromagnetic Pulses on the State's Transmission System – Interim Report*

Task Force on National and Homeland Security, comments by Dr. Peter Vincent Pry, *Rebuttal to Public Utilities Commission Report That Recommends Doing Nothing to Protect the Maine Electric Grid from Electromagnetic Pulse (EMP) and Other Threats,* December 10, 2013

EMP Coalition, letter to Maine state legislature warning against "improvements" by the Maine PUC

Cynthia Ayers, comments to MPUC, *Comments on Public Utilities Commission Report IAW Resolves 2013, Ch. 45 Relating to Geomagnetic Disturbance (GMD) and Electromagnetic Pulse (EMP)*

Chuck Manto, transcript of speaker comments from DuPont Summit, *InfraGard National EMP SIG Sessions at the Dec.6, 2013 Dupont Summit, Washington, DC,* December 6, 2013

Emprimus LLC, comments to MPUC, *Response to Draft Report on GMD/EMP Risk to Maine Power Grid,* December 18, 2013

Foundation for Resilient Societies, recommendations to Maine legislature, *Recommendations of The*

Foundation for Resilient Societies to Strengthen the Final Report of the Maine Public Utilities Commission to the Maine State Legislature on Mitigation of Geomagnetic Disturbances and Electric Magnetic Pulse Risks to the Maine Electric Grid, December 18, 2013

Center for Security Policy, statement by Frank Gaffney for the MPUC, *Statement by Frank J. Gaffney, Jr.,* December 18, 2013

Center for Security Policy, web log entry by Frank Gaffney on insights gained from the 2013 DuPont Summit of the InfraGard National Electromagnetic Pulse Special Interest Group (EMP SIG)

Federal Energy Regulatory Commission, letter to MPUC, *Re: Maine Public Utilities Commission Initiated Inquiry Into Measures To Mitigate The Effects of Geomagnetic Disturbances and Electromagnetic Pulse On The Transmission System In Maine Docket No. 2013-00415,* December 23, 2013

Civilian EMP Rating System and Sample Methods to Protect Control Systems and Networks

See document {365F6715-9070-4A28-8889-AEC2F57F0595} at https://mpuc-cms.maine.gov/CQM.Custom.WebUI/MatterFiling/MatterFilingItem.aspx?FilingSeq=78979&CaseNumber=2013-00415

Bibliography of Filings of the Maine Public Utilities Commission Docket 2013-00415 on LD131 concerning Electromagnetic Pulse and Geomagnetic Storms

Maine Public Utilities Commission, Notice of Inquiry, *Inquiry into Measures to Mitigate the Effects Of Geomagnetic Disturbances and Electromagnetic Pulse On the Transmission System in Maine,* August 21, 2013

Metatech Corp., William Radasky memo to Andrea Boland, *Role of Assessments for Geomagnetic Storm Protection,* September 4, 2013

ISO New England Inc., letter to Maine PUC, *Requested Information from ISO New England in Docket No. 2013-00415, Inquiry into Measures to Mitigate the Effects of Geomagnetic Disturbances and Electromagnetic Pulse on the Transmission System in Maine,* October 4, 2013

Emprimus LLC, memo to Maine PUC, *Emprimus LLC Answers to Maine PUC Inquiry Questions*

Foundation for Resilient Societies, filing submitted to FERC, *Comments on Reliability Standards for Geomagnetic Disturbances, Docket No. RM12-22-000,* April 1, 2013

Aon Benfield, risk management report, *Geomagnetic Storms,* January, 2013

National Defense University, Richard Andres letter to Andrea Boland summarizing research on the threat of EMP to the US electric grid, November 16, 2011

Maine House of Representatives, Andrea Boland letter to Maine PUC with linked attachments, *Comments in Support of PUC Docket 2013-00415*, October 4, 2013

Storm Analysis Consultants, John Kappenman memo to Maine PUC with embedded attachments, *Comments of John G. Kappenman, Storm Analysis Consultants*

Foundation for Resilient Societies, written testimony, *Written Testimony Before Joint Standing Committee on Energy, Utilities and Technology, State of Maine Legislature Regarding Legislative Document 131, An Act to Secure the Safety of Electrical Power Transmission Lines*, March 5, 2013

Task Force on National and Homeland Security, memo from Dr. Peter Vincent Pry to Andrea Boland, *"Quick Fix" EMP Protection for the Maine Electric Grid*, March 28, 2013

Andrea Boland, memo to Maine PUC referencing multiple supporting documents, *Comments on Maine PUC Docket 2013-00415, EMP/GMD Mitigation Study: LD 131*

Maine State Legislature, body of LD 131 plus 1 of 2 groups of written testimony

Maine State Legislature, 2 of 2 groups of LD 131 written testimony

Andrea Boland, text of speech to Maine Energy, Utilities, and Technology Committee, *Introduce LD131, An Act to Secure the Safety of Electrical Power Transmission Lines in Maine*, February 19, 2013

Central Maine Power, letter to Maine PUC with attachments, *MAINE PUBLIC UTILITIES COMMISSION, Inquiry into Measures to Mitigate the Effects Of Geomagnetic Disturbances and Electromagnetic Pulse On the Transmission System in Maine, Docket No. 2013-415*, October 4, 2013

Foundation for Resilient Societies, comments to Maine PUC, *Supplemental & Reply Comments of the Foundation for Resilient Societies Submitted to the Public Utilities Commission of the State of Maine*, October 15, 2013

Bangor Hydro Electric Company, memo to Maine PUC, *Comments of Bangor Hydro Electric Company and Maine Public Service Company*, October 4, 2013

Task Force on National and Homeland Security, Cynthia Ayers comments to Maine PUC, *Maine PUC Questions on GMD/EMP Study*, October 4, 2013

Electric Infrastructure Security Council, *The International E-Pro Report; International Elec-*

tric Grid Protection, a Report Summarizing the Status of National Electric Grid Evaluation and Protection Against Electromagnetic Threats in 11 Countries, September 2013

Andrea Boland, memo to Maine PUC, *Comments of Representative Andrea Boland,* October 15, 2013

State of Maine PUC Office of the Public Advocate, memo to Maine PUC, *Comments of the Office of the Public Advocate,* October 15, 2013

Task Force on National and Homeland Security, paper from Dr. Peter Vincent Pry to Maine PUC, *A North Korean Nuclear Pearl Harbor?*

Task Force on National and Homeland Security, paper from Dr. Peter Vincent Pry to Maine PUC, *Maine's Battle to Save America*

Task Force on National and Homeland Security, memo by Dr. Peter Vincent Pry, Dr. Fred Iklé, Professor Cynthia Ayers, Brig. Gen. Kenneth Chrosniak with foreword by Dr William R. Graham, *Civil-Military Preparedness for an Electromagnetic Pulse Catastrophe*

Maine House of Representatives, Andrea Boland letter to Maine PUC in response to the Public Advocate's comments on Docket 2013-00415, October 16, 2013

Maine Public Utilities Commission, draft report, *Draft Report in Accordance with Resolves 2013, Ch. 45 Relating to Geomagnetic Disturbance and Electromagnetic Pulse,* December 6, 2013

Ambassador Henry Cooper, memo to Maine PUC, *Comments of Ambassador Henry F. Cooper,* December 16, 2013

Andrea Boland, memo to Maine PUC, *Comments of Representative Andrea Boland in response to Draft Report,* December 18, 2013

Maine Public Utilities Commission, report, *Report to the Legislature Pursuant to Resolves 2013, Ch. 45, Regarding Geomagnetic Disturbances (GMD) and Electromagnetic Pulse (EMP),* January 20, 2014

Central Maine Power, appendix 1 of 2 to MPUC January 14, 2014 report, *GMD-EMP Risk Analysis,* January 20, 2014

NPCC Inc., appendix 2 of 2 to MPUC January 14, 2014 report, *Document C-15, Procedures for Solar Magnetic Disturbances Which Affect Electric Power Systems,* January 11, 2007

www.ingramcontent.com/pod-product-compliance
Lightning Source LLC
Chambersburg PA
CBHW051216200326
41519CB00025B/7140